Evolution, Chance, and God

Evolution, Chance, and God

Understanding the Relationship between Evolution and Religion

Brendan Sweetman

Bloomsbury Academic
An imprint of Bloomsbury Publishing Inc

BLOOMSBURY
NEW YORK • LONDON • OXFORD • NEW DELHI • SYDNEY

Bloomsbury Academic
An imprint of Bloomsbury Publishing Inc

1385 Broadway
New York
NY 10018
USA

50 Bedford Square
London
WC1B 3DP
UK

www.bloomsbury.com

BLOOMSBURY and the Diana logo are trademarks of Bloomsbury Publishing Plc

First published 2015

Library of Congress Cataloging-in-Publication Data
Sweetman, Brendan.
Evolution, chance, and God : understanding the relationship between evolution and religion / Brendan Sweetman.
pages cm
Includes bibliographical references and index.
ISBN 978-1-62892-985-0 (hardback : alk. paper) –
ISBN 978-1-62892-984-3 (pbk. : alk. paper) 1. Evolution (Biology)–Religious aspects–Christianity. 2. Creationism. 3. Evolution. I. Title.
BS651.S888 2015
215'.7–dc23
2015011137

ISBN: HB: 978-1-6289-2985-0
PB: 978-1-6289-2984-3
ePub: 978-1-6289-2986-7
ePDF: 978-1-6289-2987-4

Typeset by Integra Software Services Pvt. Ltd.
Printed and bound in the United States of America

In memory of my grandparents
Nicholas and Esther Fagan

Contents

1

Introduction: Evolution and Religion Today

Few topics concerning the general worldview of religious belief today, and especially in its relationship and interaction with science, have caused as much controversy as the topic of the relationship between religion and evolution. One only has to mention the topic in the appropriate circles to start an often contentious, sometimes unsettling, and yes, frequently heated argument. It also seems that everyone has an opinion on this topic, and, more often than not, that these opinions are strongly held, and their advocates are not usually willing to give much ground to those with whom they disagree! Unfortunately, many people adopt an entrenched position on the subject, so that there is frequently little room for informed discussion and exploration of what is a quite fascinating, but also very complex, topic. It is regrettable that there is little place in the contemporary discussion for exploring carefully and responsibly the major questions and issues involved, and their various implications for matters in religion, philosophy, ethics, science, education, and culture. The topic of religion and evolution has proved to be multifaceted, with many different themes, emphases, and perspectives, so that it becomes necessary to bring a clarity and logical order to the topic before one can begin to develop an informed view of the overall subject matter. Unfortunately, our conversation today concerning religion and evolution is frequently uninformed, especially in the public square; in addition, it is often contentious, and suffused with political implications, even agendas and ideologies, that make it difficult to bring discipline, insight, and knowledge to our thinking about this vitally important cluster of issues.

As well as attempting to introduce the general, educated reader to the main concepts, themes, and arguments of the discussion concerning religion and evolution, informed by the main perspectives in the contemporary debate, this book will consider the implications that evolution raises for philosophy, religion, and morality, and explore various ways in which evolution and religion might be compatible. It will explain the theory of evolution and the evidence for the theory, and will address a number of common questions about the theory. Then, proceeding on the assumption that the theory is true, I will argue that one of the key claims of influential evolutionary theorists—that the process contains large elements of chance and randomness—is not true. The book will discuss in detail the concepts of chance and randomness in nature, and in science, along with related theological matters, such as the question of how and why God might have created by means of evolution, what the theory means in particular for the status of human beings in creation, and questions about suffering and waste in nature. I will develop my own approach and arguments concerning these topics in the company of some influential thinkers who have written on some aspect of the relationship between religion and evolution. In this way, I hope to introduce an interesting new perspective into the debate, and so make a modest, original contribution to the overall subject.

In our discussion of the compatibility of religion and evolution, our main focus will be the religion of Christianity and the Christian understanding of God. Christianity is not only the religion we are most familiar with in the West, but it is also usually Christianity that is at the center of contemporary debates with evolution. So we will discuss some of the key questions primarily from the point of view of Christianity understood in a fairly general way, rather than in any specific denominational sense. It is the case, nevertheless, that many of the general points we make as our analysis unfolds will be applicable in many different religions if they hold that there is a transcendent supreme personal being who is responsible for creation, and also if they accept some evolutionary account of the development of life.

How to approach the topic of evolution and religion

Creationism

The topic of this book can be approached from a number of angles. One obvious starting point (especially in the United States) is the conflict between, and the controversy surrounding, what is called creationism, and the scientific theory of evolution. Many religious believers regard the theory of evolution as a threat to their (fairly literal) reading of the Bible—as contradicting this reading—and so this fact automatically puts the theory on a collision course with religion in the eyes of many of these believers. Something has to give, and from the point of view of the creationists, who generally hold that God created the universe, and especially the various species, more or less as described in the Book of Genesis, the theory of evolution has to be rejected. Of course, there are different versions of this general approach, not all of which espouse the same positions, and so we should always be careful when discussing creationism to identify correctly the specific view with which we are concerned. Some representative creationists would include Wayne Frair, Gary Patterson, Paul Nelson, and John Ashton.[1] (A caution that creationism is not to be confused with Intelligent Design theory; we will return to this distinction, and especially to the way in which the view I develop in this book differs from Intelligent Design theory, in Chapter 8.) One might be inclined to think that this creationist/evolution controversy is unique to the United States, that the rest of the world does not take this dispute seriously, that perhaps it must be something unusual in the American approach to religion that gives rise to this particular controversy. And it is true that the main reason historically in the United States for the clash between evolution and religion is that many see the theory as being on a collision course with the Bible. But this is too simplistic a way of looking at the problem for three reasons.

The first is that it is not just the creationists who are responsible for generating the controversy; many atheists and secularists were quick to jump on the theory right from the beginning and to co-opt it as an argument in support of their atheism. This fact set up a confrontation between atheism and religion, with evolution (and perhaps science in general) caught in the middle (we will come back to this point in Chapter 4). So the theory of evolution became a kind of pawn in a larger cultural debate concerning worldviews, ethics, and the meaning of life, a debate that continues and perhaps is even more intense today, as we will see in the next section. A second reason is that we do see similar problems in other countries, especially perhaps in Islamic countries, and the Far East, which generally take a dim view of evolution. The religions of India also have mixed feelings about evolution.[2] So the problem is not unique to the United States. But, thirdly, we must consider that another reason why evolution and religion do not generate as much controversy in other countries is because interested parties in those countries simply have not given enough attention to the topic. They are not as exercised, for various cultural reasons, as we are in the United States by what the theory of evolution claims, and by its implications for matters in philosophy, religion, and ethics (of course, there are millions in the United States who adopt the same attitude toward the subject!). But if educators, pastors, and academics in other counties gave more time to thinking through the topic of religion and evolution, they would have to deal with all of the questions that shape the US discussion, and it is likely that believers and intellectuals in other parts of the world would find these matters just as challenging, unsettling, and contentious as do believers and intellectuals in the United States. Indeed, if the topic of religion and evolution were to be the subject of careful scrutiny in other countries, it is likely that the theory would become more controversial in these countries, even if not perhaps to the same extent as in the United States.

It is simply a fact that whenever people think deeply about the meaning of evolution and its implications for various areas of life, including religion, ethics, and the role of science in life and culture, it inevitably raises questions *of a most foundational kind*, and these questions provoke a significant response from all sides of the spectrum. It is only because many countries do not engage in a discussion about evolution, or focus on it to any significant extent in their teaching and study of religion (and also perhaps because some maintain a careful separation between scientific matters and discussions of religion and ethics), that the theory receives less attention in these countries. It is not because people in other countries are more enlightened than their American counterparts. Enlightenment has little to do with it; the theory of evolution is complex in itself, and raises challenging questions about matters outside of evolution. These questions are not easily resolved, and require careful reflection; yet careful reflection is rarely decisive on many of the key issues, which in itself generates debate, doubt, and further controversy.

Secularism

A second approach to understanding the relationship between religion and evolution is the reverse of the first approach. I call this second approach the *secularist (or naturalistic) interpretation of evolution*, and we have already alluded to it above briefly. This approach has been enormously influential in contemporary culture, almost as influential as the presence of creationism in our culture and its various clashes with evolution. The secularist interpretation of evolution is also a significant contributory reason for why the creationist position has become so contentious, and draws so many diverse parties into the dispute (as we will see). Yet, although the secularist approach has been influential, it is not nearly as well understood as the creationist approach, so we need to take a moment here to introduce it. It very often plays an influential background role in many of the issues that arise in this book, and so it is important for us to have a grasp of this approach before taking up other crucial questions in considering the relationship between religion and evolution.

The secularist interpretation of evolution involves the manner in which atheists and secularists in contemporary society, especially those of a militant nature, have co-opted the theory of evolution (and to a lesser extent, other scientific theories) *as a way of trying to explain and justify their atheism*. And perhaps not just as a way of defending it, but as a way of trying to shore it up, give it some plausibility and respectability, render it more sophisticated, make it less negative in contrast to religious belief. Thinkers such as Richard Dawkins, Jerry Coyne, the late Carl Sagan, William Provine, and Michael Ruse would be in this camp, and I will refer to the work of these thinkers and others with similar views throughout this book.[3] Dawkins, who is probably the best known (and most notorious) representative of this approach, along with Coyne, Ruse and several others, are what I call positive atheists, or secularists. This means that they do not primarily adopt the approach of the typical atheist, who traditionally defined himself in terms of what he did not believe or in terms of what he *rejected* (God, religious tradition, religious morality, and so forth). The typical atheist of the past also *defended* his view mainly by *attacking* some other view (viz., the religious worldview), usually by critiquing arguments for the existence of God, or by pointing to the sins of religious believers, or by some combination of these kinds of arguments. However, it is obvious that from a logical point of view the alleged failure of arguments supporting religious belief and a catalogue of the all-too-real sins of religious believers does nothing to show that an atheistic, secularist worldview is true. So atheists like Dawkins, Sagan, and others came to realize that they must make a conspicuous attempt to present a more sophisticated face for the contemporary world. This takes the form of adopting a positive rather than a negative stance about what they believe, and why they believe it.

What this means in practice is that sophisticated contemporary intellectual atheists now try to think about what they believe in positive terms, to consider what they actually believe is true about reality, what really is the case out there in the real world (specifically about the universe, about human life, and about morality and politics), to express what they believe in positive statements. One might wonder how one could express in a positive way that God does not exist, or that various religious doctrines are not true, or that one does not believe in religious morality! What we need is not just some clever way of spinning these statements positively, of course, but from the philosophical and logical point of view, we need to show, if it is not true that God exists, or that human beings are designed by God, or that human beings have a soul, or that we differ in kind and not just in degree from other species, and so forth—then we need to say *what is in fact true.* And we need to work out eventually what is true on all of the ultimate questions of existence including those concerning the origin and structure of the universe, the origin of life, and the origin and nature of human life in particular, especially in relation to key human characteristics, such as consciousness, reason, free will, and morality.

Naturalists (as they are also sometimes called) or secularists[4] (I will use both terms in this book) have therefore attempted to re-express some of their views in positive statements. So they will say that what they believe is that "everything that exists is physical in nature," or that "human beings are completely physical beings, made up totally of matter and energy," or that "human beings originated as a result of an evolutionary process that operates purely by chance" or that "human consciousness is only an operation of a very sophisticated physical organ, the brain," and so forth. The point is then, according to these secularists, that if these claims are true, then their opposites would be false (i.e., various religious claims on the same topics are false); for example, it would be false then that human beings are made up of body and soul, or that human life is designed, or that human beings differ in kind from other species. By stating their views in this positive way, secularists are seen as not just rejecting or critiquing some other view (viz., the religious view), but as presenting their own independent view, which, if true, would logically entail the falsehood of the religious view.

However, in order to defend a more positive development of one's worldview or philosophy of life, one must then marshal positive arguments to support one's positive claims. We may liken the approach of contemporary atheists to supporters of the steady state theory of the origin of the universe (in the 1950s many thought this was a very plausible theory). Suppose you ask a supporter of the steady state theory why she holds the theory and she replies: "well I think the big bang theory is false." And then she offers various criticisms of the big bang theory. It is obvious that logically one cannot defend the steady state theory solely by critiquing its main rival. This is because of the logical fact that just because the big bang theory might be false, this does not show that the steady state theory is true! There will come a (logical) point where, if this scientist wishes to defend the steady state theory, she will have to give positive evidence to support the theory, a point where she can no longer rely on critiques of rival theories to establish her position. This means she will have to present positive evidence in favor of the steady state theory, not just point out how the evidence purporting to support the big bang theory does not in fact support it. It is the same with the atheist and his rejection of religion. He has an obligation, especially when he states his view positively in the way illustrated, to then present positive evidence to back up his claims.

And so it is to *science* that modern atheists turn to find their positive evidence, and especially to the theory of evolution. The great problem that this approach has given rise to (especially in the United States), and it has become a cultural problem, and not just a problem confined to the esoteric disciplines of philosophy and theology, is that these secularist thinkers often exaggerate the implications, and often the evidence, for various scientific claims and conclusions. They do this because they are attempting to give an atheistic spin to these theories, so they frequently claim more for the theories than the theories themselves can support, and even more than the scientific theories *officially claim* to support. A conspicuous recent example of this phenomenon is the claim by Stephen Hawking and Leonard Mlodinow in their book, *The Grand Design*, that the latest scientific evidence in physics, astronomy, and mathematics supports the conclusion that the universe may have arisen spontaneously out of nothing![5] They are led into making not only an illogical, but a ridiculous, claim like this by their atheistic agenda that motivates them, not only to exaggerate and misrepresent scientific evidence for their own purposes, but also, in their fervor to promote their view, to make silly claims that a moment's sober reflection by these highly qualified scientists would show are irrational!

An unfortunate consequence of this general secularist approach has been to confuse in the minds of the general public the discipline of science with atheism. When people who do not follow this complex debate too closely read thinkers (or watch them on TV) like Dawkins, Hawking and Sagan, and others, who are often public spokespersons for modern science, they come to the erroneous but understandable conclusion that modern science is really a form of atheism. When they hear scientists like Dawkins arguing that evolution should make people atheists, and urging teachers to push evolution in public education, not just to teach science, but as a way of undermining religion, then one can easily understand why they would adopt a suspicious, even hostile, attitude toward science, and especially toward programs of scientific education. Unfortunately, secularists who appeal to science in this way have fostered the view that science as a discipline (and not just science as hijacked by, and as represented by, certain thinkers with a political and moral agenda, a distinction the public does not usually make) is anti-religion, and this view has done a lot of harm to the cause of scientific education in a number of countries, especially perhaps in the United States. But from our point of view in this book, it also has the effect of making dialogue between religion and science much more difficult. Despite the championing of science by these scientific secularists, their work has the effect paradoxically of undermining people's appreciation for and understanding of various scientific theories. This leads a section of the general public to ignore, downplay, or even to scoff at excellent work in science, to weaken or even denigrate the valiant attempts by the vast majority of scientists around the world who are doing their best to understand the fascinating world around us (and, in many cases, to harness their discoveries for the benefit of mankind). These developments remind us that in any discussion of the scope of scientific theories, the evidence for them, and the relationship of science to other interesting questions and the disciplines that discuss them, it is always best to be very mindful of the distinction between science and scient*ists*!

There is a further dimension to the co-opting of cutting edge, challenging scientific theories by secularists for the purposes of advancing their moral and political worldviews. This is in addition to provoking hostility to science among the general public (and so perhaps increasing rather than reducing the level of scientific ignorance), and misrepresenting the nature of the objective debate about various scientific theories. This is the fact that some religious leaders use the ammunition (if we might put it like that) they receive from the secularist camp to avoid engaging in any attempt at constructive dialogue with scientific theories. It may be only a minority who react like this (and I want to be careful not to brand all of those religious leaders, thinkers, and believers who are more on the conservative side in the general debate between religion and science, since they are already frequently stereotyped by the secularist and liberal religious camps), but at the same time I think that secularist approaches like those of Dawkins make it easy today for any religious individual or group that wishes to avoid debate to do so. The general secularist attitude, evident in the works of evolutionary biologists like Dawkins and Coyne, or philosopher Daniel Dennett, is so hostile that it makes it easy for religious leaders to argue that these thinkers are not interested in genuine dialogue or discussion concerning matters of mutual interest in religion, secularism, and science. Religious leaders see these thinkers as being at their core anti-religious, supercilious, patronizing, and as mainly interested in advancing a moral and political agenda (under the cloak of science).

In short, our remarks in the last few paragraphs indicate that there is often a *political* side to the debate and dialogue between religion and evolution. I think it would be simply naïve today to deny this. We must be careful to keep this point in mind in our discussion to be sure that it does not interfere with our attempt at genuine dialogue, and also so that we can recognize the ways in which political matters might be influencing the debate more than we think, and how we can correct for this. We need to recognize that the political dimension of the topic plays a quite significant role in the debate that we need to be aware of; indeed, both sides contribute to it (the religious believer and the secularist), since both sides often have their own political and moral agendas. In this kind of atmosphere, it becomes easy for each side to define itself against the other, and then little genuine dialogue takes place because science gets caught in the middle, and scientific discussion may be compromised in the service of the political discussion.

Theistic evolution

We have briefly mentioned two views in the contemporary discussion of religion and evolution so far, creationism and secularism, and we will have more to say about them later in this book. But while these views often get the most attention in media discussions of this topic, because the media thrives on conflict and sensationalism, these are not the only, and not even the most interesting, positions. There are at least two other general positions on religion and evolution that are more in-between, moderate approaches when compared to the two views already mentioned. The first is the view that religion and evolution are compatible. This view is sometimes called "theistic evolution," though the term is not without controversy. It is a view promoted by a variety of mainstream religions, across many denominations, as well as by many philosophers, theologians, and scientists, including Ernan McMullin, Richard Swinburne, Paul Davies, Francis Collins, Keith Ward, and Kenneth Miller.[6] It is also the view of the Catholic Church, as noted in Pope John Paul's address to the Pontifical Academy of Sciences in 1996, where he noted that evolution is more than "an hypothesis," while stressing the indispensable role of God in creation, as well as the fact that man is made in the image and likeness of God.[7] Proponents of this view hold that our commitment to reason and science requires us to examine the evidence for various scientific theories, including evolution, and to follow it wherever it leads. If the evidence supports evolution, we must accept the theory and then need to give some thought to how evolution and religion can fit together in our overall understanding of human life. Theistic evolutionists hold that our job as interested, responsible thinkers is to explore the ways in which evolution and religion are compatible, to think about how an evolutionary account of life might fit into God's creative plan for the universe. More generally, we need to think through the various philosophical, theological, and moral implications that the theory of evolution might suggest, which as responsible thinkers we have to work through rationally and critically.

Theistic evolution does not necessarily assume that the theory of evolution is true, nor does it assume any one particular interpretation of the scientific evidence for the theory (for example, that evolution operates by chance). Those inclined toward the theory of evolution hold the view that we should examine the evidence for the theory of evolution just as we would the evidence for any other theory, without making any prior assumptions about truth or the nature of the evidence. We simply do science in this area as it is done in any other area, and rightly regard this kind of work as a branch of reason, and as part of our ongoing attempts as rational beings to understand the universe in which we live. Our prior commitment to reason, along with our increasing levels of knowledge, have already convinced us that the universe is intelligible, and so we must examine the topics of the origin and nature of species just the way we would examine any other area of nature. So if the evidence supports the theory of evolution, which many theistic evolutionists (though not all) conclude it does, then we have to accept these conclusions, and probe their implications for religious questions concerning human origins, the relation of human species to other species, and other important matters relating to the origin of life, consciousness, morality, and so forth.

It is important to emphasize that theistic evolutionists are very careful in their reading of the evidence for evolution, making sure that they are clear about what it shows and what it does not show. For example, many theistic evolutionists will disagree with the secularists that evolution shows that the development of life is governed by chance. They will dispute this, and argue that the actual scientific evidence does not show this—this is rather an extrapolation from the evidence that is often motivated by a secularist agenda. So theistic evolutionists will be extra careful to distinguish between the actual biological and other evidence, and the philosophical or religious implications that might be drawn from this evidence. This is an essential part of the approach of theistic evolutionists, and makes it unique among the four approaches we are considering here. It differs from creationism because it does not allow religious presuppositions to influence its reading of the scientific evidence; it differs from secularism because it does not allow secularist presuppositions to influence its reading of the evidence (it also differs from the next view we are about to consider in a way that we will explain as we go along). It should be clear from the remarks we have already made that a particular version of theistic evolution is the view explored and defended in this book.

Process theological approach to evolution

This brings us to a fourth response to the theory of evolution, one that is often classified as part of theistic evolution, but one that I think should be kept separate from it. This is the view that is motivated by the movement known as "process theism." Process theism is opposed to classical theism (the view of St. Augustine and St. Thomas Aquinas), and is influenced by process philosophy, a movement based on the thought of Alfred North Whitehead. The main concepts in Whitehead's thought were applied to religious topics by Charles Hartshorne, who proposed the view that reality is not static, but is in process, and that, therefore, God is best understood as having a dynamic, even organic, relationship with his creation.[8] There is a back and forward between God and the world, and God, as well as his creation, develops in this relationship. Modern proponents of this view tend to positively welcome the theory of evolution into our thinking about religion! Whereas at least some theistic evolutionists accept the theory of evolution reluctantly, and approach the topic from the point of view of showing how evolution and religion could be "compatible" or at least "not in conflict," this fourth approach, which might be described as a liberal or process theology approach to evolution (because most of its supporters are both theologically and politically liberal, or supporters of process philosophy and theology, and often both), welcomes the theory of evolution as actually making things better for religious belief in a number of ways. For instance, it would make God's creative action in history more intelligible, would make a belief in the underlying equality of species more defensible, would make modern ideas about God more plausible (such as the view that God is constantly changing through his creation). This view is promoted today by John Polkinghorne, Ian Barbour, John Haught, Arthur Peacocke, and Philip Clayton, among others.[9] Many of these thinkers also hold the view that chance can play a role in creation, but that nature overall is still purposive. (We will return to this view in Chapters 7 and 8.)

The process view also fits very well, it is thought, with modern, liberal sensibilities, such as the view that man does not occupy a privileged position in nature, that animals have inherent value in themselves (and not just in relation to us), that nature itself has inherent value, that human nature occurred by accident and so is not fixed, that the world's religions all have legitimate perspectives on reality. This position on the relationship between evolution and religion is not necessarily politically liberal, though it usually is; indeed, one must be careful again to insure that it is not a political and moral agenda for the reshaping of society and culture that motivates one's embrace of evolution, rather than one coming to believe that the evidence for evolution is strong *and* that it then supports the implications one draws from the theory. One should also be careful to insure that one's main motivation for welcoming evolution into one's theology is not because it supports a political agenda that one brings to the table independently of the evidence for evolution.

So I wish to distinguish this interesting, influential fourth approach from theistic evolution, and I call it the liberal or process view of evolution. It differs from theistic evolution because of two important points: it positively embraces evolution, whereas theistic evolutionists are generally more circumspect, holding that evolution poses some difficulties for its compatibility with theology, even if they still come to accept that the theory of evolution is true; and second, it usually goes hand in hand with a liberal theological, moral, and political agenda. Moreover, these agendas are not always easily disentangled from an appraisal of the evidence of, and general embrace of, the theory of evolution. This differs from the position of theistic evolution, where there is almost no entanglement, because theistic evolutionists try to keep an appraisal of the evidence strictly separate from their theology, and then, having studied the scientific evidence, they will try to develop the theological implications (a task they have been very slow to engage in). Whereas the liberal/process view often (but not always) simply accepts that the theory is true, and warmly welcomes it, as the starting point, and then examines its implications for theology. Sometimes these approaches, at least with regard to evolution, may appear to reflect simply a difference of emphasis, but this is not the case, since the respective theologies developed are usually quite radically different.

Evolution and contemporary culture

We have looked at four main ways that people today approach the subject of religion and evolution, but there are a number of cultural factors relating to the topic that we should also take into account. These cultural factors are sometimes influential in the overall discussion, especially at the political and social level, if not the philosophical and theological level, though these latter areas are not always immune from political and social influences. The first important point to make is that discussions involving evolution are frequently bound up with other cultural issues, which can make a dispassionate, objective, and careful analysis of the theory and its consequences very difficult to obtain. One of these cultural issues is hostility to religion, especially in academic and intellectual circles, such as in the universities, the media, education, and law. I am not referring so much to disagreement with religious beliefs, or even to holding an atheistic position, but to harboring hostility toward religion in more of an emotional rather than a rational sense. Sometimes this attitude is manifested not in a hostility toward religion in general, but in an animus toward certain expressions of religious life, and here I have particularly in mind the hostility for what is sometimes called (usually by its critics) conservative or fundamentalist Christianity (especially evident in US culture) by those who either hold more liberal religious views, or who are secularists.

The reason it is important to draw attention to this hostility—and the underlying disagreements of which it is an expression—is that discussions of, disputes about, and controversies surrounding religion and evolution are often carried out within the context of *a deeper cultural debate* involving conservative, liberal, and indeed secularist approaches to morality, politics, and the meaning of life. These disagreements also involve the secularist rejection of religion, which involves evolution being pressed into the service of atheism. So when one considers, for example, liberal criticisms of the fundamentalist view on evolution, it is important to keep in mind that they might be motivated by deeper cultural influences as well, in addition to, or even instead of, an actual philosophical, theological, and scientific appraisal of the foundational issues. The important point is that often one's position on religion and evolution is part of one's larger worldview, and so it might be that one's political and moral views play an inappropriate role in dictating one's views on religion and evolution, rather than being based on a careful study of the scientific and theological issues involved. This is why it is not uncommon to find many people today in many areas of life who have very strong views on the subject and yet who have never studied the theory of evolution, and have little or no knowledge of its philosophical or theological implications. The same point can be made about conservative critiques of liberal views, as well as of secularist critiques of religion in general. For instance, a conservative religious believer might have his own political agenda, which motivates his rejection of more liberal religious positions because of their political implications, and not as a result of any deeper theological study or considerations. Or a secularist might attack various religious views because he disagrees with these views for political reasons, yet he might utilize evolution as a smokescreen for this attack, perhaps saying publicly that he is not against religion per se, but just against the rejection of science, or against any view that mixes science and religion, or something along these lines. But a secularist in a case like this might be expediently utilizing a scientific theory in a tactical sense, as a way of giving him further ammunition against certain religious views he already rejects for political reasons. If we are honest, I think we have to acknowledge that all of these attitudes and motivations, and many more, are evident within our culture. Moreover, the philosophical, theological, and scientific issues in which we are interested in this book often get lost in the larger, contentious cultural and political debate that swirls around matters of evolution and religion.

Problems of this type are very likely to occur in a culture that is heavily politicized, such as ours. Indeed, I think it is accurate to say that many people *start* with the larger cultural debate, and never get beyond it, especially those who have little interest in the deeper questions of philosophy and theology. And our media, of course, prefers to cover the larger cultural debate rather than the substantive debate (as they do on many issues) for obvious reasons. But this only further encourages demagoguery, conflict, and political posturing and maneuvering, sows conflict, and succeeds only in confusing an already bemused public. All of this is regrettable, because it makes an objective, balanced discussion of a fascinating set of topics that much more difficult; the subject matter is already complicated enough without our allowing political and cultural issues to muddy the waters even more. So part of our aim in this book will be to keep separate the political and cultural issues from the deeper, foundational questions concerning the relationship between religion and evolution. The latter are the questions in which most fair-minded, honest inquirers are mainly interested.

Another way of expressing the above points is to say that many people often regard the question of the relationship between religion and evolution as being about a lot more than the evidence in support of a certain scientific theory, and the various implications of that theory for religion, even though these are absorbing questions in themselves, and more than enough to be going on with. Many regard the debate as bound up in different ways with larger cultural questions. It is important for us to be aware of the cultural questions as well, and how they relate to our subject, even though they are ultimately secondary to our interest in this book (and even though one could devote a whole book to this subject in itself). A particular case in point is the emergence of the Intelligent Design movement in the United States in the 1990s. This movement had a significant influence on both the philosophical and theological, as well as the cultural and political, debate concerning evolution and religion. Because the arguments of the Intelligent Design theorists extend to all of these areas, it has proved difficult to get an objective, dispassionate discussion of the claims of the Intelligent Design theorists, especially in America. Discussion of the arguments and questions of these theorists are too often bound up with other (related, but peripheral) cultural matters, such as the make up of the curriculum in public education, church versus state, conservative religious approaches versus liberal religious approaches, and religion versus secularism. (We will return to the Intelligent Design view in a later chapter.) Our point is not that these larger cultural questions are not important, but that they are *not* the foundational questions; these latter questions are our concern in this book.

The compatibility of religion and evolution

The aim of this book then is to explore various ways in which evolution and religion may be compatible with each other. The book is written with the aim of being accessible to the general, educated reader who has an interest in the topic, and is not meant only for specialists in philosophy, theology, or science. We will approach our topic by arguing for a particular view on religion and evolution, specifically with regard to the question of chance in nature. Central to my thinking will be the claim that, contrary to one influential, even prevailing, view in the general discussion, evolution does *not* operate by or involve a significant element of chance. This means that there is no chance operating in either physics or biology. I will defend, with certain qualifications concerning human free will and any action by God in creation, the position of determinism with regard to the operations of the physical universe. I will also suggest that there is a clear sense in which evolution is moving toward goals, a conclusion that also applies to the whole of nature (and therefore to the whole of science). This position then provides us with a new foundation from which to argue that evolution and religion are compatible. I have become convinced in my reading and thinking about evolution that, with certain qualifications, there is no chance or random occurrences involved in the process, and that leading thinkers in biology, philosophy, and theology who claim there is, or who assume there is, or who have made chance a key part of the way they understand and explain the theory, are seriously mistaken. Moreover, this mistake has far-reaching consequences for the evolution/religion discussion. I will explore the implications of this determinist position with regard to causal happenings throughout the universe for evolution and religion, especially with regard to the question of their compatibility.

My overall position in this book is that God exists, and that the general religious perspective on reality is a very powerful one. I will assume this position in this book. The general religious view of the world is not only enormously satisfying from a psychological and emotional point of view, but I regard it as clearly rationally superior to any secularist alternative. (I will elaborate on this claim in Chapters 7 and 8.) It offers human beings a rational hope in the ultimate meaning of existence, and makes sense out of many features of our universe, including its value-laden nature. As philosopher Nicholas Wolterstorff and many others have pointed out, human beings are naturally religious,[10] and so in a sense a religious view of reality is really the default view, and will likely remain so. So, if we accept that God exists and that evolution is true, then not only is it true that evolution is part of God's creation, but we would have to conclude that it is an important part of that creation. And so perhaps we need not ask so much *if* religion and evolution are compatible, but should move on to ask *how* they might be compatible. We will explore how the compatibility of religion and evolution might be best understood by means of an exploration of the question of chance in nature, and also of possible suggestions and scenarios for how we should understand their actual relationship. Along the way I will examine the views of various thinkers on these questions, but will be mainly interested in developing my own explorations and suggestions for how evolution and religion can go together in an attempt to understand how nature works as part of God's creation. Many affirm that evolution and religion are compatible, but my concern in this part of my discussion will be to probe our thinking more in terms of specifics about how the theory of evolution and religion could be reconciled. This will enable us to address more deeply than we normally do key foundational questions concerning chance, design, our place in nature, suffering, waste in nature, morality, and God's way of acting in and through creation.

There are many books on the subject of religion and evolution, from many different perspectives, but there are none that I know of that make the concepts of chance, randomness, necessity, determinism, predictability, and probability in relation to the overall discussion a central focus in the way that we will attempt in this book. And there are none that explore the question in detail of how religion and evolution might (or might not) be compatible from a philosophical point of view. There are some very interesting books that explore some of these matters from a theological perspective, but they assume much more than one can assume in a philosophical discussion, and they are also quite speculative. This is particularly true of the approaches found in process theology.[11] There are also some that start from a position of secularism, and that generally dismiss the question about the compatibility of science and religion, and that fail to discuss the key topic of chance and necessity in nature in anything other than a superficial way.[12] And there are still others—perhaps the majority on the religious side—that hold that evolution and religion are compatible, but that do not probe or explore in depth the ways in which they are, or could be said to be, compatible. Moreover, in many of these studies there is an assumption that evolution operates by chance; this is not just a standard line in biology, one now finds it too often in theology and philosophy. A central argument of my work on religion and evolution involves calling this common understanding of the role of chance in evolution into question. In calling it into question, I will not argue so much that God must have designed evolution, but rather that when one looks at the theory of evolution purely as a scientific theory—and the nature of the universe more generally (and science, and the scientific method)—the evidence shows that there is no chance operating throughout these processes. One implication of this finding then is that the process of evolution might suggest design after all, *and* this conclusion has very significant consequences for our understanding of the implications raised by the theory of evolution.

I am approaching the issues from the point of view of assuming that evolution is true, though the early part of the book will provide a full discussion of the theory, including the evidence for evolution, and questions related to the evidence. It is necessary to understand what we mean by the theory of evolution, and how it is generally defended, before we can address adequately its implications for matters in philosophy, religion, and morality. But my approach to the topic is this: let us say that we accept that the main claims of the theory of evolution are true, and we also believe that God exists, then it must be the case that God created by means of, and indeed intended, the process of evolution. Therefore we must, in our philosophical and theological thinking about religion, explore why God created this way, as well as exploring the implications of evolution for other matters, such as the place of human beings in nature and in relation to other species, the apparent randomness of evolution, the presence of much disease, evil, and hardship in nature, the future of evolution (say in 100,000 or in a 1,000,000 years), evolution and morality, and the role of chance in causation. I wish to explore in some detail, and from a philosophical point of view primarily, how these issues might be developed, *given* that we accept that belief in God is reasonable, and that belief in evolution is reasonable.

It seems to me that, at least when considered on the surface, or intuitively, many people, both theists and atheists, think that evolution is a problem for religion. We might say that this is the conventional wisdom about their relationship. This is partly because there are sets of theists and atheists who focus almost completely on the problems in the relationship, and perhaps emphasize and/or exaggerate the difficulties at the expense of any genuine dialogue. But even many theologians think there may be a problem; as we noted above, they reluctantly accept evolution and try to "accommodate" it within religion in various ways. Yet I believe that a change in attitude is called for, and may be fruitful in trying to go forward on the issue, and I am commending that change in attitude to readers of this book. This change in attitude will require all of us to put the "conflict" model of religion and evolution completely behind us. It will require us to approach the matter from the point of view that evolution does not operate by chance (or at least that whether it does or not is an *open* question); it will also allow us to have an honest and interesting discussion about the state of the evidence for evolution, and what it shows and does not show. It also asks us to detach ourselves from the wider cultural debate, and, while being aware of it, to leave it to one side for the purposes of what is already an absorbing discussion. This goes for atheists as well as religious believers. I am not saying that this will be easy for any of us to do, for the influence of the conflict model and the cultural matters that swirl around evolution are deep and pervasive in our thinking about these topics. Nevertheless, two things can be said here: we must have a change of attitude if we are to make progress on this matter; and second, the *foundational* questions are still there whether we address them adequately or not. So we must address them; nothing less is required from our general commitment to reason and evidence. I also predict that in the coming decades the change in attitude that I am recommending here will become more widespread in this discussion.

The dialogue model

Our approach in this book is motivated by the dialogue model of the relationship between religion and science in general, and also specifically in the area of evolution. The dialogue model is committed to several key positions. The first is the traditional claim of Christian philosophy, supported by both St. Augustine and St. Thomas Aquinas, that "all truth is one." This is a common sense, logical commitment to the position that if a claim is true, or a fact has been established, or a theory proved to be true in one discipline, or area of life, it must be true in all disciplines. Nothing else is acceptable if we are committed to reason and logic. For example, if the big bang theory is true in science, then it must also be true in theology (and every other discipline). It is also true in political theory, economics, and psychology, but, of course, would not usually be relevant in these disciplines; in short, it is true *objectively*. The reverse is true as well—if a theory or a claim is true in philosophy and theology, then it must also be true in science. For example, if it is true that God created the universe, or that a human person consists of body and soul, then these claims are also true in science. This is not to say that science has to be able to prove these claims; it only means that if it is a fact that God created the universe, then this fact would be true in all disciplines of study, including science, whether or not the discipline in question ever considered the matter, ever took the fact into account in its specific disciplinary work, or even whether the practitioners of the discipline believed the claim. But it would also then follow that science would not be able to disprove this fact (if it is a fact).

Our second position applies this first one to religion and evolution. Our stance is that belief in God is a reasonable belief, as is the belief that God created the universe and all life for a purpose. I will not defend these claims in this book but will adopt them as our working hypothesis in our discussion of religion and evolution. But I do want to emphasize that I am holding that the religious view of the world, including the above claims, is rational, and more rational than any atheist or secularist alternative. Given this, it is my view that we must examine the evidence for evolutionary theory just as we would examine the evidence for any other scientific theory. As thinkers committed to logic and reason, and evidence, nothing less is required of us. As noted above, we will also proceed from the assumption that evolution is true and work within this assumption, though I do devote two chapters to a full exploration of the theory, because every reader today who is interested in these topics and who wishes to explore their implications must be informed about the theory in some detail. It is not enough to have only a superficial knowledge.

So given that God exists and created the universe and all life, and given that evolution is true, then it is simply a fact that evolution and religion are compatible. Our job is to think about how they are compatible, no matter how things might look on the surface. We need to ask some difficult questions: about how evolution is part of God's design, about why God might create through evolution, about what the actual process of evolution is meant to convey to us, about the significance of the fact that beings have evolved who understand the theory, about evolution and chance, about evolution and evil, about evolution and morality. Along the way, we will examine the views of important thinkers who have taken interesting positions on some of these questions. My approach will be mainly one of exploration, aimed at stimulating further reflection, but I will also tentatively advance my own position, especially with regard to issues concerning chance, randomness, determinism, and necessity in nature.

Our approach is in contrast to two other models regarding the relationship between religion and science in general, and to the relationship between religion and evolution in particular. These are the "independence" model and the already mentioned "conflict model." The latter model presents the relationship as being one of conflict, and people on both sides of the topic regrettably adopt this model. But this model does not accurately represent the historical relationship between religion and science, since they usually worked quite well together (with occasional exceptions such as the Galileo affair in the sixteenth century, and the evolution/creationism and the secularism/religion controversy today). It also does not represent the correct relationship between them today since I am claiming that the respective truth claims of religion and evolutionary theory are not, and cannot be, in conflict, no matter how things may appear on the surface. But the conflict model gets much media attention, thereby setting back in a very serious way the possibility of progress, not just in terms of lessening conflict, but in terms of understanding the correct relationship between religion and evolution.

It might seem that the independence model is a better model to adopt in this debate than the other two. Proponents of this model, who include theologian Karl Barth and evolutionary biologist Stephen J. Gould, advocate keeping religion and science in strictly separate domains where religion would keep to religious topics, and science would deal with a study of the natural world, but crucially we would not worry about whether scientific and religious claims appear to clash with each other.[13] According to this view, religion and science deal with different subject matters, and it is best we keep them distinct for a number of reasons. The domains have different methodologies, and each methodology is not appropriate in the other discipline; they have different concerns and questions that do not overlap—religion with ultimate questions of meaning, morality, and salvation; science with how things work, causes and effects, the makeup of physical objects, etc. There would be added benefits too to keeping them separate in the present cultural climate that should not be underestimated: one is that people would not have to worry about whether their work on, or interest in, or understanding of, science created problems for religious belief, and vice versa; another would be that over time this approach would lead to less hostility toward science among the general public that would translate into more support for, and so an improvement in, science education in our schools. Indeed, it might be argued by proponents of the independence model that by keeping them separate it would isolate more the secularist position that misuses and even abuses science to further an ideological cause, and that this would be a step forward in the cultural debate.

While these are all very interesting points, we do need to consider whether the independence model is a logical approach to the question of how science and religion might be related. More specifically, it is interesting to ask whether proponents of this model are denying the idea that "all truth is one," mentioned above. Are they saying that if we find something to be true in theology, then it is possible for a scientific theory to come up with a different truth, and both disciplines could be right? For example, suppose that we conclude in the discipline of biology that those who say that the human species came about by chance are right (that this is what the evidence likely shows), and suppose we hold in our religious worldview that the human species was designed by God, are proponents of the independence model claiming that both of these positions are true? This kind of problem forces us to ask whether the independence model is recommending that we ignore contradictory truth claims from different disciplines, for the purposes of dealing with these problems better from a psychological and cultural point of view, or is it making the stronger (and more relativistic) claim that truth is relative to each discipline? It seems that these thinkers are making the stronger claim because the first claim would seem to amount to a dodge and would not honestly engage or explain the difficulties that we need to deal with whenever truth claims in different disciplines appear to clash. The independence model could be understood to be a psychological way of ignoring a logical problem, but the logical problem is still there. This approach also makes it difficult for a Christian who is both a scientist and a believer to integrate her life, because it seems she would need to adopt an almost schizophrenic approach of being a Christian in her life but a scientist in her work.[14] Yet the alternative understanding of this model is just as problematic because it would be committing us to a relativistic interpretation of truth, since it would be advocating that contradictory claims in different disciplines could all be correct. So the problem with the independence model is that, while it may promote a superficial harmony, it appears to do so at the expense of logic and truth. In our attempt to reach an overall understanding of reality, we must utilize all of the best reasoning and evidence we have from all disciplines, and this is why the dialogue model is superior to the conflict model and the independence model.

A philosophical approach

As I have indicated, our approach in this book will be philosophical, and not theological, secularist (atheistic), or scientific (though I will discuss all of these approaches along the way as well.) The discipline of philosophy can be very helpful in sorting through the many complex issues raised in any study of religion and evolution. Indeed, it is regrettable that more philosophers have not brought their expertise to the discussion (except as cheerleaders for the conflict model); but I want to emphasize before we bring this chapter to a close the central importance of a philosophical approach to our subject. I think it is fair to say that the academic discussion on evolution and religion in the contemporary world has suffered from a lack of attention by philosophers; the conversation is usually dominated by theologians and secularists (some of whom are also scientists), and so the discussion often lacks the clarity, rigor, and focus on foundational questions that philosophy brings to a subject. (The contemporary discussion also suffers from a lack of participation by scientists, especially by those who are not atheists or motivated by any kind of anti-religious agenda, a group that contains the vast majority of scientists, of course.) The philosophical perspective is a valuable perspective from which to approach a study of religion and evolution, and indeed religion and science in general. Indeed, I think that this debate badly needs the philosophical perspective, because we have been poorly served by those atheistic scientists, in particular, who have addressed the issues with superficial analyses of very important questions and occasionally with prejudices toward philosophy, metaphysics, and theology (they are particularly weak when it comes to questions dealing with necessity, chance, laws of nature, probability, predictability, teleology, and the problem of evil).

The discipline of philosophy is concerned with some of the most foundational questions of life. The approach of the philosopher involves thinking through these questions by considering the views of pivotal thinkers, presenting and critically analyzing the reasons offered for various views, considering objections, and subjecting one's overall position to rigorous critical analysis. Philosophers also seek ultimate causes or explanations, rather than local causes or explanations. The local cause or explanation would be the one that various areas of inquiry in their everyday practice would seek; for example, we might conclude that the earth formed out of the debris of matter left over in space from the big bang, or that poverty plays a causal role in crime. But philosophers ask more fundamental questions about the ultimate *cause* of the earth, the deeper cause, why we ended up with any earth, or universe, at all, or about why evil exists in the world, and so forth. Although answers to ultimate questions stretch our minds as far as they can go, the questions must still be asked; otherwise we are leaving out something fundamental.

Philosophers focus on logical argument, and on identifying important distinctions, in order to bring clarity to complex issues. But philosophers do not overlook empirical evidence if it is relevant to the questions in which they are interested. We have argued that the principle that "all truth is one" requires us to consider all of the evidence wherever it may lead, and especially its implications for theories and claims in any discipline. Philosophers differ from theologians, in particular, in an important way: they try not to assume anything that is too controversial in their arguments; philosophers try to defend their assumptions as much as possible. Theologians can usually assume the reasonability of a particular religious tradition, and the reliability of certain religious texts, and then do excellent work given these assumptions. Philosophers like to look at more foundational questions, and are required to defend their various answers to these questions by appeal to reason, human experience, and philosophical arguments. But as I have already noted, we will operate with some basic assumptions for our purposes in this study, the main one being that it is rational to believe in God, and that God created the universe and all life. But it is important to point out that that is only an assumption for our purposes here; the philosopher would be required to defend these claims elsewhere, and then could assume them in considering the question of the relationship between religion and evolution.[15] We can assume them here because our aim is to explore ways in which religion and evolution might be said to be compatible, rather than *if* they are compatible. That is not our topic in this book.

Our next chapter begins with an overview of the discipline of science, its methodology, and the type of (objective) knowledge that science seeks, and includes a brief reflection on the nature of scientific theories. Then we turn to a detailed introduction to the theory of evolution: its historical background, main claims, including an overview of its central concepts (natural selection, survival of the fittest, microevolution, macroevolution, common descent, and so forth). The aim is to explain the main claims and themes of the theory for the general educated reader as clearly as possible so that we all know what we are talking about in the evolution/religion discussion, and so that one can appreciate later the various implications that it is often argued the theory has for various religious and philosophical matters.

Chapter 3 turns to a consideration of the evidence for evolution, presented in a question-and-answer format. Questions considered include: what is the evidence that natural selection occurs (here I will argue that it is best to make a distinction between two kinds of natural selection when thinking about evolution); what is the evidence to support common descent; what is the DNA evidence; is evolution only a theory, and can evolution be falsified? Our aim is to summarize as objectively as possible the evidence that is currently offered to support the claims of evolution by leading thinkers, and to identify any weaknesses or gaps in the evidence, and critical questions that one might ask about the evidence. Although for the purposes of our discussion in this book we accept that the theory is well supported by the evidence, it is also important to consider any weaknesses or pressure points in the evidence, and what (more generally) is still unsettled with regard to the theory and its main claims.

We turn in Chapter 4 to identify some of the reactions to, and implications of, the theory for matters outside of science, such as in religion, philosophy, and ethics. After surveying the historical and contemporary reaction to the theory, we ask such questions as: is evolution an ideology; are there different types of evolution; what are the theological implications of evolution, especially for our understanding of the Christian Bible; what are the philosophical implications, such as for the view that human beings differ in kind and not just in degree from other species, what are the implications with regard to the concepts of chance, necessity, and determinism, and the role they play in the operation of the universe?

Chapters 5 and 6 turn to the heart of our argument in this book. Chapter 5 considers a key argument appealed to by a number of evolutionary thinkers that the process of evolution contains a large element of chance and randomness, and that this conclusion undermines in a number of ways the idea that there is any design in nature. I will argue that there is no chance operating in the universe at any level, that the view that evolution is governed by chance is a very significant confusion in evolutionary biology. This confusion has significant consequences for our understanding of the implications of evolution for matters outside of biology, such as in philosophy and theology. This chapter first illustrates the concepts of chance and randomness as they are used in evolutionary biology in the work of leading thinkers, including Stephen J. Gould, Richard Dawkins, Elliot Sober, Ernst Mayr, and Jacques Monod. The notions of chance, randomness, and determinism are then considered more generally as they apply to the universe as a whole, and in the discipline of physics. Chapter 6 then develops the argument presented with particular reference to how the main concepts of chance and randomness would apply in the process of evolution, especially with reference to the phenomenon of mutational changes in DNA. This view is illustrated by reference to the examples used in the work of Mayr, Gould, Monod, and Jerry Coyne. The distinction between chance and randomness, and probability and predictability, is then clearly explained, and several confusions identified in the application of these concepts to evolution. The last part of the chapter takes up consideration of an objection to my deterministic view of nature from within quantum mechanics, and also considers the question as to whether or not the process of evolution is progressive.

In Chapter 7 we turn to the implications of our previous analysis for how we should understand some key concepts in religious belief, especially the notions of design, creation, and God's plan for the universe. We discuss various ways that religion and evolution are compatible, and respond to some of the implications of evolution for religion, raised in Chapter 4. After identifying a serious problem for naturalistic accounts of reality, the chapter goes on to consider the role of divine action in creation. Why would God use a process like evolution to create the species; does the process undermine the uniqueness of man; does the evidence show that God might be directing evolution, or that evolution is not directed; does God "intervene" specially in creation in some way, or simply allow it to evolve; what roles do the concepts of primary and secondary causation play in creation; could God allow some role for chance in nature after all? Several different views will be considered, including some proposals from recent work in theology by John Polkinghorne and Arthur Peacocke.

The final chapter continues to develop the themes of Chapter 7, especially with regard to the place of design and chance in nature, and in the universe. The chapter considers whether there could be a mixture of chance and determinism in creation, and how these concepts relate to the question of the overall design of the universe. I also distinguish my view from the position of Intelligent Design theory. We then turn to the implications of my deterministic thesis for the question as to whether evolution can be considered as an argument against design and God because it is a wasteful, inefficient, even cruel, process, and so it makes the problem of evil worse than it was before evolution. I offer a critique of this argument. The chapter concludes by considering some of the problems that evolutionary naturalism would present for secularist accounts of morality. These include the problem of justifying the objective moral order through evolution alone, the related problem of relativism, the difficulty of justifying Christian moral actions such as altruism and self-sacrifice from an evolutionary standpoint, and the dilemma of explaining the phenomena of free will and moral agency in general.

This book develops and brings to fruition an argument I have been thinking about for a number of years, and that has been hinted at and sketched out briefly in some of my earlier writings. I have sometimes used a phrase or two, or a couple of paragraphs, from material I have published previously; in addition, parts of Chapters 5 and 8 are refinements of earlier arguments. I wish to thank Bloomsbury, the editors of the *American Catholic Philosophical Quarterly*, and of the *Portuguese Journal of Philosophy*, for granting permission. Although they do not necessarily agree with all of my arguments, I am grateful for stimulating conversations on the topics of this book with friends and colleagues, as well as with the participants at conferences and symposia in Europe and America, especially: Paul Allen, Bruce Ballard, Richard Carlson, Fr. Gabriele Gionti, S.J. (of the Vatican Observatory), Doug Geivett, Catherine Green, Curtis Hancock, Fr. Bill LaCroix, S.J., Emily Mazzoni, John Morris, Santiago Sia, Brendan Sweetman, Jnr., Mindy Walker, and the late Ernan McMullin. I owe a special debt of gratitude to Bill Stancil for his erudition and depth of theological knowledge, for his incisive comments on the whole manuscript, and for his enthusiastic support of the project. I am grateful also to the anonymous referees for Bloomsbury for helpful comments on several draft chapters. I also wish to acknowledge Rockhurst University for providing me with a summer research grant. I am grateful to the excellent staff at Bloomsbury Publishing, most especially to Haaris Naqvi, for their professionalism and hard work. My greatest gratitude is reserved for my family for their always unfailing support and encouragement, without which this book would not have been possible!

2

The Theory of Evolution

Before we can examine the implications of the theory of evolution for various matters in religion, philosophy, and morality, and especially in relation to the role of chance and necessity in nature, we need to have a good working understanding of the theory. It is essential today to have an informed view on what we all know is a quite complex topic. Unfortunately, but perhaps understandably, there are many who do not have a sufficient understanding of evolution to make informed comment on the issues of the debate. Yet, this lack of knowledge does not always prevent them from making a contribution anyway, which just adds to the confusion! There are also many scientists who would only have a passing knowledge of the theory of evolution. This is not to make a criticism of these scientists; science is now a vast area of knowledge, and the discipline is becoming increasingly specialized. Many scientists likely only have time to concentrate on their own areas of research, and might settle for a passing knowledge of other areas (as scholars do in most other disciplines). There are also many religious believers who would have a limited understanding of the theory, including theologians and religious thinkers, but again this does not always hinder them from speaking out on the theory, either for or against it, depending on their perspective! So it is important to have a good working knowledge of the theory if one wishes to think more clearly about the questions it raises, and how to respond to these questions. We can here only provide a summary of the theory for the general, educated reader, not for the specialist. The debate over the fine points of detail in the theory of evolution is fascinating in itself, and is carried out by some very smart minds. But an overview of the main themes in evolution is sufficient for our purposes in this book.

Three important questions to ask about the theory of evolution

I like to approach the theory of evolution from the point of view of three main questions. The first question concerns the theses of evolution: what are the *main claims* of the theory; what does the theory claim to be true about its subject matter—the origin, nature of, and development of, the various life-forms, organisms, and species that live or have lived on earth; for example, what does the theory say about the relationship between fish and birds? The second question concerns the evidence: what is the *evidence* offered to support the main claims of the theory, and is it good evidence; for example, what evidence would show that modern elephants are descendants of elephants that lived millions of years ago? The third question turns to the *implications* of evolution: what are the implications of the theory for topics outside of science, especially in the areas of philosophy, religion, and morality; for example, if it is true that chimpanzees and human beings are related to each other, what might this imply, if anything, about the uniqueness of man in creation; what might it imply for the biblical story of creation? We will discuss the first question in this chapter, the second question in Chapters 3 and 4, and the third question in the remaining chapters of the book. I think these three questions help us to organize the subject matter of evolution, and keep the topic as clear as possible. And although in this book our focus is primarily on the third question, it is essential to have a good knowledge of the first two questions before we can consider the third question in an adequate way. While our primary concern is not with whether or not the evidence (question 2) does support the claims of the theory of evolution (question 1), we will provide a full discussion of some of the common sense questions that interested readers will have about the evidence and what it shows. This discussion about the evidence is not only very interesting in itself, but it will give us a solid background and allow us to examine in a more insightful way the many questions that the theory raises for religious matters. It is important to have a good general understanding of evolution in order to address some of the deeper philosophical and theological questions it raises.

Some preliminary points about science and the scientific method

Before we get into the theory itself, we need to say a little more about three positions that are usually presupposed in a discussion of any scientific theory. These positions concern the nature of science, the scientific method, and the type of knowledge at which science aims. There is often uncertainty concerning these matters that can then lead to misunderstandings when we try to examine the implications of scientific theories. These misunderstandings, as we will see, are most common among scientists, but they sometimes appear in the work of theologians as well.

First, the discipline of science involves building up a store of knowledge about the physical world. This knowledge is important for two reasons. The first is an intrinsic reason, which means that as human beings we simply want to have knowledge of how things work for its own sake. For example, we want to know the composition of the big bang, how light travels through space, the causes of the Ice Age, the effect of heat on subatomic particles, the mechanisms involved in blood circulation, how a virus causes a certain type of disease in humans, and so forth. This research fulfills our natural curiosity about and craving for understanding of the world around us. But, second, of course there is also an instrumental reason for knowing these things. The knowledge gained may help us improve our lives in a variety of areas (for example, we might be able to develop a cure for the virus). This kind of scientific work is based on observations of the natural world in various areas, the gathering of data, the carrying out of experiments, and common sense reasoning about all of these matters. Occasionally, hypothesizing and formulating theories might play a role at any point in the process. A scientist might formulate a theory in advance (which is what happened in the case of the study of immunity when German scientist Paul Ehrlich carried out numerous, varied experiments in an attempt to discover the causal processes behind it[1]), and then try to test the theory by experiment and further deduction (this was also Einstein's approach with the theory of relativity). Or it might be the case that a scientist assembles a set of facts and other empirical data first, and then proposes a theory to best explain the facts and data (which was Darwin's approach with the theory of evolution).

Second, the scientific method has become the primary method for studying the realm of the physical. In general, we understand this realm much better than we did in the past, and we now tend to distinguish it and perhaps separate it off completely from the realm of the spiritual and the supernatural (at least in academic circles), something that was not always done in the history of the discipline, even by scientists themselves. But increasing scientific analysis of the physical realm has confirmed the theory (first proposed by the ancient Greeks!) that the realm of the physical is made up of particles in motion, and involves relationships between matter and energy. We know from scientific work in the twentieth century that this realm is made up of atoms that are related in various ways, and also of various subatomic particles (particles smaller than atoms, of which atoms consist). Although the scientific method can sometimes appear (and indeed often is) quite complex, inaccessible, and even mysterious to those with little or no knowledge of science, it actually involves pain-staking, tedious, time-consuming, often frustrating (and expensive) attempts to carry out experiments, gather data, propose and test hypotheses, debate rival hypotheses, and arrive at testable conclusions that are fairly well supported by reason and evidence.

Scientists are also assisted, especially today, in gathering evidence by the use of increasingly sophisticated technology; indeed technology often develops as scientific knowledge develops. Moreover, the development of technology is itself a scientific enterprise (e.g., building the space shuttle and the Hubble telescope); then the technology enables us to apply the scientific method to discover more knowledge about the physical realm (e.g., studying far off galaxies by means of the Hubble telescope). So we can say that the scientific method involves making observations, gathering facts and data, and performing experiments, all using the very latest technology available. In addition, it also involves forming hypotheses, theories, and explanations in order to explain various phenomena, and then testing them, where possible. We must also note that studying the physical realm is now such a large endeavor, that science has increasingly become more and more specialized, as scientists devote their time to studying only little parts of reality (e.g., the nature of the atom, the ecology of the rain forest, a particular species of plant, and so forth). This is part of the reason for why scientists often do not know much about science outside of their own area, and why many know little or nothing about the theory of evolution simply as a scientific theory (i.e., they don't know much more about what the theory claims or about the evidence for the theory beyond what the general public would know), let alone about its implications for other topics.

It is important, third, to discuss briefly before we come to the theory of evolution the nature of the knowledge that science seeks. There are a number of important sub-points that we need to keep in mind as we begin to think about one of the most significant scientific theories of all time. The first is that science deals with facts and theories, and the search for objective knowledge. To elaborate this point, we can distinguish in science between a scientific theory and a scientific claim (or what is sometimes called an observation claim).[2] An observation claim is one where scientists "observe," record, or verify some event that is taking place in the ordinary world of our experience, some *fact* that they note as true. For example, they might observe that a certain plant grows well in a certain type of soil, or they might report that a certain type of wood rots easily when subjected to wet conditions, or that a certain type of substance is strong enough to carry (say in a hose) high-pressure liquids for a considerable period of time, etc. The everyday business of science works with these types of claims. These are claims that involve an appeal to reason and sense experience; at some level, we must be able to observe and touch the data that constitutes empirical evidence in science. Other scientists (and, in principle, anyone) should be able to observe the same phenomena, and their interactions and effects, and be able to follow the process of reasoning proposed concerning any particular scientific claim or hypothesis—this is the process of verifying the data in a public, universal way. This general process is what is meant by "empirical evidence" in science.

The second sub-point to note is that these facts scientists observe are proposed as objectively true facts. Such facts are supposed to report the way the world really is; they are describing nature as it really is independent of the human mind, and especially independent of our opinions, biases, hopes, and wishes. For instance, when scientists argue that a code in a person's DNA leads to the production of proteins, and that one group of these proteins carries oxygen in our blood, they are saying that this process really occurs in the body (meaning that a certain combination of DNA code will lead to the production of a protein of a certain type and that this protein will then carry oxygen). It happens irrespective of our political, moral, or religious beliefs, irrespective of its implications for any of these matters. It is simply a fact. But, of course, third, scientists often *disagree* about what the facts show because the evidence may be sparse, sketchy, imprecise, inconclusive, or hard to obtain. Some scientists, for example, may disagree that the evidence about various gases and the ozone layer shows that the former will harm the latter. While there is often disagreement about what the facts show, we must note nevertheless that whoever asserts a claim as a fact is saying that *this is the way the world really is*. Unfortunately, there is no clear line of demonstration in science, or any other area of knowledge, between asserting that something is a fact and its actually being a fact, between claiming that something is true and its actually being true. While there is a clear line in logic, of course, it is not so easy to maintain the line in practice because we decide whether something is true in science usually when there is a *consensus* of scientists who agree that it is true. And we accept that when this happens, the likely reason they think something is true is because it is true (i.e., they are guided by the evidence as reasonable people trying to understand things in the best way they can).

Yet we know that logically scientists could all think that something is true that is not, in fact, true. This is an important point because we are aware that the history of science is full of examples of scientists accepting that a theory is true, and then later coming to see that they were wrong. For example, in the late nineteenth century scientists posited that there was a luminiferous ether-like substance around the earth that light traveled through, but the existence of the ether could never be satisfactorily confirmed by experiments. Einstein's theory of relativity was regarded as a decisive argument against the ether theory, and it was later abandoned. But these erroneous claims in science are usually regarded as honest mistakes. The common sense answer to the question about the status of scientific claims and theories from the vast majority of scientists is that they are put forward as objectively true claims, as telling us about what really happened, about how things are really related or structured, about how the world really is. Indeed, this is the common sense view of this matter in all areas of reasoning, not just in science. Historians, psychologists, theologians, and sociologists are all making factual claims about reality. The main point is that it is important to remind ourselves that while the discipline of science employs reason and common sense, and the scientific method, in an attempt to understand the realm of the physical, and while it seeks out the facts and the objective truth about this realm, it is sometimes hard to get agreement about the facts, and the discipline sometimes makes mistakes.

This brings us to our third sub-point, which concerns scientific theories. There is a difference between scientific theories and ordinary factual or observation claims. Scientific theories involve the more theoretical part of science, a part in which the average scientist often plays no role. Major theories in science only come along occasionally, and often bring fame to the scientists who propose them; they are then debated, applied, and utilized by other scientists, as well as by other concerned intellectuals outside of science. Of course, there are many lower level breakthroughs in many sub-disciplines of science (e.g., the discovery of the combustion engine), but these seldom involve major theoretical work at the foundational level the way the theory of relativity, or evolution, or the laws of motion, gravity or thermodynamics would. Some of the major theories generate lively debate, disagreement, even extreme controversy (e.g., evolution, the big bang theory, the theory of global warming). There are many reasons for why a theory may generate controversy. A main one is that on the purely scientific level, scientists will propose a theory to explain the facts they see before them. The theory should explain the facts, of course, yet it is often the case that the theory goes beyond the facts as well. The nature of science (and indeed human reasoning in general) is that scientists will sometimes propose conclusions that are not fully obvious given the facts before us, or will offer conclusions that are suggested by the facts, but not proved by them. A theory may stick closely to the facts; but some theories might be more speculative, and therefore more subject to debate. A theory gains respectability, as we have noted, if there is a general consensus among scientists and other thinkers that the facts upon which it is based are true, that they do support logically the theory that is proposed to explain these facts (though we have also noted that a consensus by itself does not guarantee that a theory is true, logically).

This leads us to our fourth and last sub-point. Sometimes a theory is controversial because of its perceived implications for other areas outside of the theory. This is the case with evolution (and it was also the case with heliocentricism). Evolution has proved controversial because many think that it has significant implications for religious issues. But our central point is that we must be careful to distinguish between what a theory claims is objectively true, and *the implications that are then drawn from these objective truths.* We do not always maintain this distinction in the complicated and often messy business of trying to figure out what is factually true in different areas of inquiry. We must be careful especially in the discussion about evolution to keep separate what the theory actually says is true in the world from the implications that might be derived from these truths. This means that there are several things we must be very careful to avoid: we must avoid claiming that evolution explains things that it does not explain; we must avoid downplaying the evidence for evolution because we worry about its implications; we must avoid exaggerating the evidence because we want the theory to be true; we must avoid allowing our own worldview or philosophy of life (especially our political and moral beliefs) to influence our interpretation of the theory, or our evaluation of the evidence; we must avoid confusing the political and moral views of a particular scientist with his or her scientific work. These are all common problems in the discussion concerning evolution and religion, and we will come back to them occasionally in this book. But it is important to emphasize at the beginning that we must take special care to avoid all of these problems that have hampered reaching a proper dialogue and debate on the theory, and that have also hampered our attempts to understand and address more fully the deep issues at stake.

Historical background to evolution

The theory of evolution was developed by Charles Darwin (1809–1882), an English naturalist, in his book *The Origin of Species*, which first appeared in 1859.[3] In 1831, Darwin was chosen as a scientist for the British ship, HMS *Beagle*, which embarked on a five-year round-the-world trip. It was his research on this voyage that began to suggest some of the ideas that were to become the foundation for his later theory. The ship's itinerary included a trip to South America, where Darwin visited the Galapagos Islands in 1835 (these islands are 600 miles off the coast of Ecuador). Darwin worked fairly quietly after returning from the voyage at his home in Downe, near London in the United Kingdom, working primarily on developing his ideas about evolution, and especially in trying to gather enough evidence to support what he knew was a radical theory. He wanted his ideas to appear plausible and convincing, and worthy of being taken seriously by his fellow scientists. Darwin never held a position as a university professor, and seldom gave lectures or public talks. He had a small circle of mainly scientist friends, and only confided the main ideas of his theory to a limited few, though apparently many knew he was working on something original in relation to the vexing question in biology of the origin, nature, and development of species. This topic had interested and puzzled those in the newly growing field of biology for some years. Before Darwin, there was no accepted theory concerning the question of the origination of new species. Many scientists accepted some version of the view often called "the fixity of the species," the belief that the species existed in their present form, and that very little change in them had occurred over time. It was still unclear to many scientists how new species came into being; there was also a lot of doubt about whether extinction of species was possible or had occurred, about the age of the earth, and related topics. Many held that the species were created just as they are by God, a view that was consistent with the Bible and also with Aristotle's physics, which was a dominant philosophical approach in the understanding of nature.

Darwin was also aware of the fact that his theory would have some relevance for religious belief. He was especially mindful of the arguments that had been put forward by English theologian, William Paley, in his well-known and influential work, *Natural Theology, or Evidences of the Existence and Attributes of the Deity* (1802).[4] Paley was a strong supporter of natural theology, the attempt to examine any evidence for the existence of God that can be found in the physical universe (including from the study of life), and then attempting to draw some conclusion about whether or not God's existence can be inferred on the basis of the evidence found. Natural theology has usually reached a positive conclusion on this matter. In his *Natural Theology*, Paley relied primarily on the design argument for the existence of God, an argument that in its general form says that there are various evidences of design in nature, and this design then suggests that there must be an Intelligent Mind, namely God, behind the universe, just as, to use an analogy, the design we put into nature (e.g., in a watch) bespeaks our intelligence. However, what was interesting about Paley's approach was that he often appealed to detailed descriptions of physiological aspects of animals (e.g., the intricate structure of the eye) and habitats in nature as examples of the kind of design that we find in the world. This type of design had long fascinated many, including biologists. But Darwin recognized that his theory would offer an alternative explanation for these phenomena in nature. His theory would describe a completely natural process that brought about the nature and structure of various species (and perhaps it could even be applied to the habitats as well). So even though he was simply describing what he believed had taken place in nature, and trying to present the evidence that led him to that conclusion, he saw that, indirectly, his theory could also be read as a refutation of Paley's version of the design argument (an argument Darwin had initially been convinced by).[5] He also knew that some would likely interpret it that way, and that some would question his motives.

It is also true that some thinkers welcomed Darwin's theory because they liked the fact that it could be employed as an indirect attack on that form of the design argument proposed by Paley. Darwin realized that in publishing his theory, it would be very difficult to keep the science and the theological implications separate, and that he would be stepping into a minefield. Indeed, his own religious views are often the subject of some interest, with many wondering whether his religious beliefs were influenced negatively by his views on evolution. This does not seem to have been the case, and here and there in his papers he does raise occasionally some of the interesting questions evolution presents us with, but does not discuss them in any detail, or show much evidence of having thought about them deeply (e.g., he once noted that chance was just another name for our ignorance of causes, but did not go on to develop the implications of this understanding of chance for the claim that evolution might operate by chance). And let us not overlook the fact that several times throughout the *Origin* he refers to the Creator, and different ways that the Creator might have brought species into being.[6] It seems as if Darwin did come to doubt his religious beliefs at the end of his life. Yet, this appears to have been because of his worry about the presence of evil in the world (brought home to him by the premature death of his daughter from illness) than because of his theory (though because of evolution he became more aware of the presence of evil and suffering in nature in general, a point we will return to later in this book).

The Origin of Species is concerned with providing an account of how all of the various species we see in nature, including not just human beings but also animals, insects, even plants, came to be. Darwin was not so much interested in what I called in Chapter 1 the "ultimate question" of how they came to be, but in the more localized question of how they came to exist in nature when they did, how they came to have the physical makeup and structure they have, what their ancestral lineage might be, and so forth. We know from intellectual history that it is rare for a scientist, or any other thinker, to discover a theory completely on his or her own, with no help from previous work by other dedicated scholars, and Darwin's theory is no exception. Several of the most significant concepts that Darwin brought together in proposing his theory had been individually developed before him, but their full significance had not yet dawned on earlier scientists. Indeed at one point another scientist, Alfred Russell Wallace, wrote to Darwin (in 1858) and expressed very similar ideas to the ones Darwin was working on, causing some later discussion about who had discovered the main ideas first. Other influential forerunners and contemporaries of Darwin were James Hutton (1726–1797), George Cuvier (1769–1862), Charles Lyell (1797–1875), Carl Linnaeus (1707–1778), and Jean Baptiste Lamarck (1744–1829).[7]

Cuvier, a French naturalist, is often credited with developing the fields of comparative anatomy and paleontology (the study of life in prehistoric times) because he carried out a systematic study and classification of fossils. Fossils are the skeletal or trace remains of animals and plants that have been preserved in various ways in clay, mud, rocks, and other strata, sometimes for billions of years. Most fossils are excavated from sedimentary rock. In studying and classifying fossils, Cuvier started the process of focusing on the origin, habitats, and lifestyles of various species. This involved studying their anatomical structures, attempting to relate them to other species, and learning something about how they lived, including about their likely biological structures, diet, prey, predators, and environmental conditions. He is known for the theory of catastrophism, the view that the earth had been hit over time by a series of catastrophes (most likely floods) that were responsible for wiping out various species, and which also significantly affected the environments of surviving species. The theory of catastrophism might also be an explanation for the significant gaps in the fossil record. Cuvier's discoveries included the classification of various species of elephants; the proposal (later confirmed) that there was a time when reptiles had been dominant; his principle of correlation of parts—he argued that by examining a single bone of a species found in a fossil, one could reconstruct the species, and sometimes the genus, to which it belonged, a good insight into how comparative anatomy works (and also of its speculative nature). But Cuvier rejected evolution because his research into fossils was unable to document significant changes over time in the fossil record, and so he thought that species bred within their type, with only small variations.

Jean Baptiste Lamarck, another French naturalist, was among the first to present a more developed model of evolution, and his views had a significant influence on Darwin. He is famous for an alternative theory of how evolution worked, often called "the inheritance of acquired characteristics." Lamarck argued that the environment plays a large role in the development of the biological structure of an organism. Examples he gave to illustrate were the way the repeated hammering by the blacksmith leads to the development of bigger bicep muscles in his arms over time, and giraffes having to constantly stretch into tall trees for food, an action that gradually leads to the development of longer necks. The idea is that this repeated behavior gradually leads to the development of the various characteristics in species along certain lines that are advantageous to the species, and that help the species to survive. Lamarck then argued that these "acquired characteristics" would be passed on to the offspring of the species in question, so that, to return to our giraffe example, eventually most giraffes would have long necks. His proposal for how various species obtained their particular characteristics was intriguing, and many were inclined to accept his views initially (including Darwin). Lamarck also held that evolution was gradual, that the species were not fixed, but could change over time. Some scientists, including Cuvier, rejected the inheritance of acquired characteristics on the grounds that these characteristics were often not passed on in future generations.

A pioneering thinker in the early story of evolution was Charles Lyell, an English geologist, who published his very influential *Principles of Geology* in 1833. Lyell was influenced by geologist James Hutton's gradualist view that the current state of the earth could be understood by studying past causal activities, and that great change was brought about by small successive changes over a very long period of time. Lyell thought it reasonable to assume that the scientific laws that he observed operating in his own work had always operated, and so he came to the conclusion that the state of the earth in various fields, including geology and biology, was shaped by the laws of science operating on prior causal states over very long periods of time to produce present states and events. This view was known as uniformitarianism, and was presented in opposition to catastrophism, though the ideas are *not* incompatible once one remembers that catastrophes would be produced by the laws of science like any other event, and their effects would also occur according to the laws of science with the catastrophe being the prior cause. The basis of Lyell's understanding of physics and how the physical laws of the universe would affect biology will be relevant later in our discussions in Chapters 5 and 6 about whether or not chance exists in nature. Lyell's basic approach assumes that no chance operates in nature, and his idea that we could understand current biological outcomes by looking at past biological and geological states had a strong influence on Darwin.[8]

The theory of evolution: Key concepts

Darwin was struggling with the question of the origin of, and structure of, the various species we see in nature, such as finches, pigeons, bats, squirrels, and dogs. He was concerned with common sense questions that many interested parties were asking, such as: how did a new species come about? Is there a clear difference between a species and a variety? How should we classify various animals, insects, even plants? Are varieties that are similar to each other thereby related to each other (e.g., varieties of dogs), and by extension are similar species related to each other (e.g., different species of birds)? What about species that are more distant from each other—might they also be related (e.g., chimpanzees and human beings)? Did species develop over time, or were they fixed in their structure and nature? We have already noted some of these fascinating and very difficult questions above, and noted briefly some answers to them (e.g., Lamarck's). Over the course of about twenty-five years, Darwin developed his own theory to answer these questions; it was different from, and indeed a rejection of, previous theories, and it also had radical implications for matters outside of biology.

One of the things that Darwin's *The Origin of Species* did was that it brought together a wide array of facts and research in biology and geology arranged around a single organizing theory: evolution. This was one of the features that made his theory attractive. It also meant that piecing together the evidence to support the theory would be an important task. One can approach evolution as a theory in two ways. The first is to consider a vast array of facts and data (e.g., concerning fossils, habitats, predatory behavior, diets, the various characteristics of species, observation of species change) and then to think of evolution as a theory that one brings to or applies to the facts and data, that helps one to interpret the facts and data, and in the light of which the facts and data make sense (for example, the theory might explain why the forelimbs of humans and birds are similar in structure). The second is to say that the facts *point to* the theory—are evidence that this is the way species change happens; the facts confirm that something like this theory is true (e.g., DNA evidence shows that all life forms are likely related). While a theory can be approached in either way, and indeed most theories work from both directions—they attempt to explain the facts well (top–down), and the facts suggest and support them (bottom–up), it is not always clear in the theory of evolution which of these approaches should be emphasized over the other. Evolutionary biologists would likely argue that it works in both directions equally well and equally effectively for evolution, just as it does for many other theories (while critics would argue that if a theory relies too much on the top–down approach, this in general is a weakness in a theory, and that this is the case with evolution).

It is helpful to begin a more detailed exposition of the main points of the theory of evolution by listing the key concepts.[9] These concepts are: (i) there is change over time among species and varieties (sometimes called adaptation of the species); (ii) natural selection; (iii) the struggle for existence and the survival of the fittest; (iv) microevolution; (v) macroevolution (sometimes called common descent, common ancestry or descent with modification); (vi) the progressive nature of evolution (this means that, in general, the process of evolutionary change starts with simpler life forms and species, and gradually leads over long periods of time to the emergence of more complex life forms and species). These themes are interrelated and overlap in various ways, as we will see. There are also a number of subsidiary themes that are not really part of the theory, but that often come up in association with it. We will address these in our later discussion. These subsidiary themes would include: the origin of life; the origin of the universe; the underlying laws of the universe; the evolution of consciousness and related features of human life such as reason and logic, free will and moral agency; evolution as an organizing principle of science (not only biology); and the role of chance in evolution.

Before we begin to explain the main concepts of the theory of evolution, we need to say something about the concepts of species, variety, and other types of classification used in biology. Carl Linnaeus was the first to develop a detailed theory of species classification (influenced by Aristotle's classification), and his system is still used today. Linnaeus decided upon species as the basic unit of classification for living things (including plants), and then introduced higher orders of classification. These were subsequently added to and refined by later thinkers. The basic structure of classification today in biology begins with species (and within species there are varieties); then we have genus, family, order, class, phylum, and kingdom, in increasing order of generality. The general way of classifying living things has been influenced by the thesis of common ancestry, and the hierarchy inherent in these groupings naturally suggests (though is not of course evidence for) common descent. This order of classification also has affinities with the tree of life, which is another way of illustrating in diagrammatical form the relationships between all living things. A popular example used to illustrate the order of classification is the species of cat. The ordinary house cat belongs to one species of cat, and the mountain lion belongs to another, but both are in the same genus because they are types of cat (which means they have some similar characteristics, e.g., four legs, similar body structure, claws). But not all cats are in the same genus; lions and tigers are put in a different genus because they differ enough from the other two to warrant a separate class (so it is argued). The cheetah is in yet another genus, and all three of these genera would then be in a family of cats, the next general category. Cats in general and dogs in general are placed together in the next class, which is called Order, and so on up the table. (Plants are classified using the same taxonomy.) One can see that this is a very useful way of classifying various animals and plants, but that it might not be without controversy as to where a species is classified, though, despite this, the basic structure is sound and very practical.

The concept of species is the central unit of classification in this system. It refers to life forms of a certain general type or kind, and then the higher order categories include more general types that seem to be of the same kind (i.e., lions and tigers can be classed as separate species, and then higher up the table of classification they can be put in the same genus, as we have seen). So to be more specific then, a species is usually defined as a group that interbreeds among themselves, and produces fertile offspring, but that does not breed with any outside group; this is sometimes described in biology by saying that a species is a group that is "reproductively isolated" from another group (e.g., western, eastern, and mountain gorillas are now regarded as different species, but previously they were thought to be the same species).[10] A genus groups together species of the same type (for example, all the various species of gorilla). There are also varieties within a species; these would have slight variations in their characteristics, for example, (two) varieties of western gorilla, the different varieties of finch Darwin saw on the Galapagos Islands, the many different varieties of dogs, etc. The definition of species is not without controversy, of course. It is somewhat fluid, and it can be difficult to decide on important points relating to whether some species can interbreed but just don't, or whether some varieties are actually different species, or even when a new species begins and an older one ends (or breaks off or becomes extinct), as well as about how to classify species, more generally.

But the basic thinking is that we have some living things that clearly belong in the same species (e.g., polar bears); some that belong to the same species but are different varieties because they have slightly different characteristics (for example, greyhounds and bulldogs); some species that are quite different yet somehow related (for example, wolves and dogs); some that are not closely related and that clearly belong in very different species (e.g., dogs and horses); and some that are widely different from each other (e.g., fish and birds, or apes and humans). All species are related, of course, according to the thesis of common ancestry, yet in these latter cases the relation is not clear or obvious, as it is in some of the earlier cases, and so would require stronger evidence, and might generate debate and controversy. Linnaeus and others thought that the best way to take account of most of these differences in classification is in terms of whether or not living things can interbreed with each other. This is the basic notion behind the definition of a species.

Illustrations of natural selection: Finches and bacteria

So keeping all of these points in mind, one of the most helpful ways to illustrate the main concepts of the theory is to work through a few examples. Biologists usually turn to a few favored cases to illustrate the main concepts of the theory. Evolution is often illustrated with what I call fictional examples (what Darwin called "imaginary illustrations"[11]), as well as by appeal to real cases from the world of nature. Imaginary cases are often better for the purposes of illustration because many of the details can be made up and filled in to show what the theory is claiming, and to illustrate the main concepts of the theory; the problem with the real cases is that we have to reconstruct so much of the story that it often appears very speculative, and not fully convincing, and this creates obvious problems. Evolution is a fascinating theory; it is easier to explain and illustrate its main claims; it is more difficult to produce strong evidence to illustrate them. Among the standard examples cited to support the theory are the finches Darwin studied on the Galapagos Islands, and bacteria in the human body and their reaction to drugs. Let us further illustrate by discussing these two cases.

Darwin was fascinated by the variety of flora and fauna he observed on the Galapagos Islands; he also noted that many of the species that lived on the islands were different from any he knew about elsewhere in the world, though he noted that many species native to the islands had similarities to some of those on the mainland.[12] This was especially true of the various types of finches he observed on the Galapagos. He speculated that millions of years earlier mainland finches had flown out to the Galapagos, and that the present finches on the islands were the descendants of the mainland ones. The mainland finches were of a different species than the present Galapagos species. They may have had different colors, different body sizes, different wingspans, and, in particular, different beak sizes. Darwin did not know how the original finches obtained these various characteristics (genetics was unknown at this time), though he believed this was not a crucial matter for his theory; the important fact is that they have these various features, and that they are passed on to the offspring by the parents. He proposed that over very long periods of time, using the finches on the Galapagos as an example, there was a struggle for existence going on in nature between the various species, and that those who survived this struggle did so because they had some kind of advantage over other species. So those finches, for instance, that survive do so because they are the "fittest," meaning not the strongest or healthiest, but that they are best able to cope—apparently because of chance—with the particular environment they find themselves in. The finches that flew out to the Galapagos Islands found themselves in a new, different environment and gradually adapted to that environment.

One of the main differences Darwin observed between the different species of finch, and various other birds, was their beak size. He argued that the beaks of the finches (to take a specific biological characteristic) evolved in various ways due to environmental factors that influenced the source and supply of food. He noted that one species had a large beak that was necessary for cracking open the seeds it needed to eat to survive. Another species fed on passing insects; it had a large beak that made it easy to catch insects, and he speculated that the beak had adapted to the conditions over time. He speculated that the finches that flew out to the islands originally that did not have these characteristics eventually died off. Sooner or later, those finches with beaks suitable to find the available food came to predominate in a certain area, and to interbreed, and passed on their characteristics to their offspring. So only finches with beaks suitable for eating the type of food available in that area survived. And so he rejected Lamarck's view that the repeated beak-activity of a finch would lead to the development of a longer beak. What happens, he argued, is that by chance some have long beaks and some have short beaks to begin with, and those with long beaks survive better, and pass on their characteristics (including the long beak) to their offspring, so that eventually long-beaked finches dominate. This process of adaptation to certain environmental conditions he called natural selection. It is the mechanism or process of change by which the characteristics of a species like a finch can change over time. As Darwin put it: "Owning to this struggle for life, any variation, however slight and from whatever cause proceeding, if it be in any degree profitable to an individual of a species ... will tend to the preservation of that individual and will generally be inherited by its offspring ... I have called this principle ... Natural Selection, in order to mark its relation to man's power of selection."[13] The idea of natural selection was suggested to him by his study of artificial selection widely used by farmers in England to produce desirable herds of cattle and dogs. If artificial selection could be used to produce small changes in animals in a relatively short time, then natural selection might produce significant changes operating over a very long time. Darwin got the idea of survival of the fittest after he read (in 1838) a pamphlet by English cleric and economist, Thomas Malthus, in which Malthus argued that since there are limited resources, there would inevitably be a struggle for existence among human beings. Darwin reasoned that nature, with its cutthroat competition for survival between different species, could be understood in the same way.

Let us turn now to the example of the development of antibiotic resistance in bacteria to illustrate further the thesis of natural selection, since it is such a central concept in evolution. Bacteria are useful to study from the point of view of evolution because they exist in huge numbers and multiply rapidly, so that a mutation that makes a species of bacteria resistant to a new drug is likely to occur given enough time, which in the case of bacteria can be as little as a month.[14] Influenced by the work of pioneering thinker in genetics, Gregor Mendel, scientists posited the existence of genes in various species to explain the passing on by parents to offspring of traits that later turned out to be advantageous for survival (recall that Darwin was not sure how traits were passed on; he just knew they were passed on to offspring). Mutations involve changes in the DNA of the organism. The unraveling of the nature of DNA itself was a breakthrough in the study of genes, and indirectly in the development of thinking about evolution. Alterations in the genes of an individual—or mutations—would affect the traits being passed on to the offspring of an organism, and hence their survival rates. These mutations are caused by problems inherent in the DNA itself, as well as by a variety of often hard to document environmental factors, including diet. So the existence of genes that are subject to mutation, and the process of natural selection, would work together to produce new species. This process is evident in the study of bacteria and their reaction to drugs in various medical treatments. The process is simple to understand, and is offered as evidence of the process of natural selection in action. (We will discuss the concept of mutation in detail in Chapters 5 and 6.)

Suppose a patient is suffering from a blood disorder that is caused by a variety of bacteria. The patient takes an antibiotic to kill the bacteria. But suppose that *some* of the bacteria (but not all) have a genetic makeup that prevents them from being killed by the antibiotic the first time the patient takes it. This means that the chemicals in the antibiotic are not sufficient to kill off completely this particular strain of bacteria. Many of the other strains that contribute to the blood disorder will be killed off by the drug, and the patient may have a temporary reprieve from the illness. But, according to the thesis of natural selection, the strain that is resistant will begin to multiply and will eventually become dominant in the blood. The second time the patient takes the drug, it will kill off other vulnerable strains that are also present, but not the continually multiplying resistant strain, and the drug will have less effect than the first time. This pattern will be repeated until the patient has in his body only the resistant strain, and then the drug will have no effect. This is an example of natural selection in action: the resistant strain of bacteria already had a trait that makes them initially resistant and they passed this on to their offspring, and gradually this strain began to dominate while the others died off. If there is no resistant strain at the beginning of the process, one may soon appear because of the speed at which bacteria multiply.

This is an interesting example of natural selection, yet it is often misunderstood. It is important to note that the critical strain of bacteria was either *originally* resistant and this is why it survived, or that a resistant strain emerged because of the speed with which bacteria reproduce. A strain did not develop resistance as a reaction to the drug, meaning that one particular type of bacteria did not suddenly develop a section of chemical code in DNA that it did not have originally so as to be able to cope with the drug. It just happened *by chance* to develop this section of chemical code that *luckily* enabled it to fend off the antibiotics (at least according to contemporary interpretations of evolutionary theory). So when we say that the bacteria evolved, we mean that one species of bacteria had a chemical code that enabled it to dominate over the others in the way explained; we do not mean that this species developed a chemical structure that it originally did not have. This misunderstanding of the way evolution works is very common, with many thinking "the bacteria evolve new traits to make them resistant."

Microevolution and macroevolution

The concept of microevolution is what is being illustrated in these examples. Microevolution is defined as evolution within a species, or within and between closely related species, such as within the class of finches, or elephants, or whales, or in our last example, bacteria. One of the key claims of microevolution is that species *are related to each other genetically*. This means that the finches that Darwin studied on the Galapagos Islands and the finches that he studied on the mainland are related to each other genetically in that they had common ancestors, viz. the finches that originally lived on the mainland and later flew out to the islands. Adaptation and natural selection, he argued, could explain why the finches on the island changed over the years in some of their characteristics, giving rise to new varieties and indeed to new species. The varieties would be finches that have some differences, e.g., different beak sizes, colors, perhaps wingspans, but which still interbreed together. The new species would be finches with sufficient differences so that they no longer interbreed together; they are likely geographically separated as well. The same process would take place with the mainland finches; they too would be different today in some respects from their ancestors, but the key point is that the mainland finches and the island finches would have common ancestors, if one goes back far enough.

So a key thesis of the theory of evolution is this thesis of common ancestry or common descent (Darwin called it descent with modification).

Darwin extended the idea of natural selection to argue that it could explain not just the beak sizes or body sizes of certain birds and animals, but that it could explain *all* of an organism's features.[15] For example, it could explain why certain birds have a certain type of body size, wing span, eye structure, hearing apparatus, heart setup, lung capacity, skeletal arrangement, and so on. It could explain why species' eyes have the structure they have; how their heart and lungs (and ears, nose, throat, and indeed any feature you care to mention) came to have their particular structures and configurations. It is difficult to explain all of these features in the details, since all of the fossils are mostly lost, and we do not have any detailed knowledge of the actual development of particular life forms and species. Nor do we have precise knowledge of the environments in which the various species lived, nor of the causal conditions to which all of these features were subject. Yet natural selection can explain their development in the general sense that each part of each organ evolved so as to confer some selective advantage on the organism that aided its survival (usually over its competitors).

Darwin next proposed the thesis of macroevolution. To return to our finches' example, his theory raised the question as to whether similar species of finches were related to each other, as we have seen. He concluded they were. This then gave rise to the question, are *all* species of finch in the world genetically related to each other? Might they all have the same common ancestor? Might they all be descendants from the same ancestral species, or line of species? His answer to this question is also yes. Then this latter question gave rise to an even bigger question: might it be the case that widely different species (in different Genera, different Orders, different Families, etc.), such as human beings and chimpanzees, are genetically related? Might fish and birds have a common ancestor? For all that, might human beings and chimps, and fish and birds, have a common ancestor? His answer here again is yes. This then prompts us to ask if it might be the case that all the species of animals and insects that have ever lived are genetically related to each other? And a yet further question is whether all animals and all plants might have a common ancestor? And finally, whether all life forms might be traceable back to a single life form that came into existence billions of years ago? Darwin answered all of these questions in the affirmative, and sketched out a tree-like diagram in his book illustrating what the thesis of common descent might look when applied to all living things.[16]

So Darwin's theory of evolution proposed the radical claim that all living things and species that live now or that have ever lived are genetically connected to each other, including not just human beings and lower animals, but also insects, bacteria, other simpler organisms, and even plants. This genetic connectedness of all the species is nowadays known as macroevolution. Darwin came to the view that all of the present species in the world evolved from common ancestors, right back to the very first life forms, which some speculate were one-celled organisms that appeared around 4 billion years ago (in the sea). He argued that more complex life forms emerge over time from simple life forms, by means of the mechanism of natural selection—the struggle for survival in nature in which only the fittest survive, and which also determines the physical makeup of the various species.

Timeline of evolution

A diagram of the history of the whole process of evolution (which can be found in many biology textbooks) is referred to as the tree of life. In most books, it begins with one single-celled organism, which reflects the view of many evolutionary biologists, though some argue that life might have begun with separate single-celled organisms in different parts of the world. The timeline proposed for the evolution of life is as follows. The universe itself, according to the big bang theory, is 15–20 billion years old. Our galaxy/earth is approximately 4.5–5 billion years old. Current evidence suggests that life began on earth around 3.5–4 billion years ago. We do not know how it began, and officially the theory of evolution is not concerned with the origin of life, but only with the origin and development of species after life appears, but some do try to apply the theory to the origin of life itself (as we will see in Chapter 4). The earliest evidence of life is found in fossils that are dated to be more than 3 billion years old. These fossils were found to contain cells known as prokaryotic cells, a fairly simple bacteria-type cell that does not have a nucleus. For the next 2 billion years, this is the only type of life found in fossils. Then in fossils dated to be about 1.5 billion years old, we find the first instances of cells known as eukaryotic cells. These are cells that have a nucleus surrounded by a membrane that contains DNA.

A study of the fossil record shows gradual evolution of life in more complex ways from around 700 million years ago (progressivism of evolution), when we begin to see multicellular organisms appear for the first time. Around about 520 million years ago there occurs what is now called the Cambrian period, in which we find the "Cambrian explosion," the rapid appearance of many major species ancestral to several modern species (such as shellfish and corals). The best known source of fossils in this period is the Burgess Shale, discovered in the Canadian Rockies in 1909. By the end of the Cambrian period, there is evidence for nearly all major animal groups.[17] After this point, new organisms appear in a pattern of historical succession that is very well documented in the fossil record. The first fish appeared in the Ordovician period (480 million years), the first amphibians appeared 380 million years ago, the first reptiles 40 million years later, and the first dinosaurs appeared 80 million years after that in the Triassic period (250–200 million years ago). The first birds appear in the Jurassic period (155 million years ago). All of these conclusions are based on a study of, and interpretation of, the fossil record.[18] Only in the last 3 or 4 million years did the ancestors of human beings and the higher primates begin to appear.

The fossil record appears to show that human beings and chimpanzees had a common ancestor (6 million years ago). Gradually two distinct lines of new genera split off from this common ancestor, evolving through a few million years so that eventually we have new species of chimps on one line, and of *Homo sapiens* (human beings) on the other line, with many related species in between. For example, on the line that led to *Homo sapiens*, we also had earlier species, such as *A.africanus* (3.5 million years ago, of which the female skeleton, Lucy, is the most famous fossil); *H.habilis* (2 million years ago); *H.erectus* and *H.neanderthalensis* (Neanderthal man), our closest ancestors (500,000–2 million years ago), and then *Homo sapiens*, the only surviving species from this lineage (200,000 years ago). Modern humans are thought to have originated in a line of descent from these earlier species about 60,000 years ago. This timeline clearly shows the progressive nature of the process of evolution in that it starts with very simple life forms (prokaryotes) and gradually over billions of years leads to the emergence of very complex life forms that have consciousness, rationality, and moral agency. The attempt to document all of these claims is obviously very difficult, ongoing, and a fascinating journey in itself. And the whole matter takes on a much larger significance when individual claims concerning the discovery and dating of a fossil, its individual makeup, its classification, similarity to other species, and speculation about its lineage, are all considered in relation to the larger issue of evolution, especially the concepts of macroevolution and natural selection.

Carl Sagan used the clever metaphor of a "cosmic calendar" to bring a sense of perspective to the timeline involved not only in evolution, but also in the origin and development of the universe.[19] In this calendar the history of the universe is compressed into a calendar year. Some of the significant dates would be: January 1st: the big bang; May 1st: formation of the Milky Way (our galaxy); September 14th: formation of the earth; September 25th: appearance of life on earth; December 24th: first dinosaurs; December 30th:

first hominids; December 31st: first humans. A month in the cosmic calendar represents one and a quarter billion years, a day 40 million years, and a second 500 years. An average human life lasts less than one second in the cosmic calendar year, which gives some sense of the vast timeline involved in creation!

So as we conclude our overview of the main claims of the theory of evolution, we should emphasize its two key themes: the notion of macroevolution and the thesis of natural selection. Macroevolution is the view that new (radically different) species come about through natural selection, and that all of the species are genetically connected to each other in the way just explained (Darwin called this connectedness "descent with modification"). Natural selection describes the process of change by which one species gradually evolves into a different species, and perhaps we should describe it also as the process by which new species come into being. Both of these (obviously related) claims proved hugely controversial, and both had massive implications for topics outside of science. We must also underscore one very large point: the theory of evolution is not an atheistic theory in itself. It is an attempt to explain the processes in nature that led to the origin of species, and a study of these processes also led Darwin to the conclusion of common descent. But the theory makes no claims about whether or not the evolutionary process was set in motion by a designer, nor about whether God might be directing the mechanisms of evolution in an overall sense. It makes no further claims about whether evolution can explain the origin of matter, or the laws of the universe. Indeed, how could it, since these questions are outside of its subject matter, and scientific methodology. However, in some of its twentieth-century versions, it is not so easy to disentangle these other issues from the theory itself, as we will see. Yet it is always crucial to distinguish between the theory and its possible implications for larger questions. Before we turn to these implications in later chapters, we must first consider our second question concerning the evidence that is offered to support the main claims of the theory of evolution, and this will also give us a further opportunity to elaborate a bit more on the main concepts of the theory.

3

Evolution and the Evidence: Questions and Answers

Too many discussions of evolution seem to gloss over the question of the evidence for the theory. No doubt this reluctance is partly because of the cultural climate surrounding the theory. It sometimes seems as if those who work in the area of evolution are hesitant to discuss the evidence in an open, honest, serious way in case they give ammunition to their creationist critics; some religious believers perhaps don't want to look closely at the evidence because they have already made up their minds. This is unfortunate because the question of the evidence is obviously a very important one, and my own view is that it is a quite fascinating topic in itself (leaving aside all ideological, philosophical, theological, and moral implications). It is also, of course, a large question, and it has to be said, something of an elusive question, which is another reason why it can be sometimes difficult to get a clear perspective on how reliable the evidence is. The evidence for the theory is such an interesting, but controversial, matter that it deserves a chapter onto itself; indeed, we must be mindful of the difference between giving an explanation of the theory—explaining how the theory is to be understood (which we did in the previous chapter)—and showing that the theory is actually true, or at least that it is quite reasonable to give one's assent to the main claims of the theory.

We should also point out again that, while our aim in this book is not to undermine evolution, but instead to probe its implications for other important matters on the assumption that it is true, and to examine how evolution and religion might be compatible, it is also obviously essential that we have an understanding of the main lines of evidence offered for the theory. We need some appreciation for whether the evidence is strong or weak, or somewhere in between. We also need some understanding of the deeper questions that might be raised by the way the evidence is often presented (e.g., what role does chance play in the explanation of the theory; is evolution progressive?). In this chapter, we will provide an overview of the main lines of support offered for the theory, consider some objections, and raise other issues that are germane to the discussion of the evidence, such as whether it is accurate or fair to describe evolution as "only a theory," and whether it would ever be possible to falsify evolution. I will adopt a question-and-answer format when discussing the evidence. I find this a helpful, fruitful strategy for elucidating the evidence and arguments used to support evolution. The questions raised are the ones the general educated reader asks about the theory, and responding to these questions will enable us to explain in a lucid way what the evidence is, how it fits together, what the weaknesses might be, and how the evidence might raise other lines of inquiry that relate to the implications of the theory. (The implications will be then brought out in the next chapter.)

Two kinds of natural selection: Is the evidence convincing?

We noted in the previous chapter that natural selection is the process or mechanism by which changes take place in what are called varieties within species, and in species themselves (microevolution). It is a process that eventually produces dramatic changes that lead to widely different species (macroevolution). So while it is true that the thesis of natural selection and the thesis of common descent are two different, separate theses, they are nevertheless connected in an intimate way. This is because natural selection is the process by which not just microevolution but also macroevolution occurs. And macroevolution means that all of the various species, both now and in history (including insects, all the lower life-forms, and plants), are genetically related. Natural selection explains the process by which a species of finch with predominantly short beaks evolves into a species of finch with predominantly long beaks over time. But it is also supposed to explain the process by which the descendants of fish gradually over a very long period of time evolved into birds. It also describes the process by which the species of great ape gradually evolved into the species of *Homo sapiens*, again over a very long period of time (at least 20 million years).

What happens is that the great ape, from which *Homo sapiens* has descended, initially evolved into a species that was different from, but similar to, the great ape. And this species evolved again into one that had further differences. This process occurred thousands of times, right down to the emergence of *Homo sapiens*. Along the way, many of the intermediate species became extinct, including all of the ancestors of man; for example, *A.africanus* (who lived 3.5 million years ago), *H.habilis* (2 million years ago), *H.erectus* (1.8 million years ago), and *H.neanderthalensis* (Neanderthal man, 100,000 years ago) are all extinct species. The process driving change from one species to the next, each change small in itself, but adding up to dramatic change over time, is natural selection. It is important to note that one could still accept this common descent from ape to man without believing in natural selection; yet without natural selection we would need to discover some other process that explains and drives evolutionary change. The lines of evidence for common descent and for natural selection seem to be independent of each other—so that if we doubted natural selection, the independent evidence for common descent (from the fossil record and DNA evidence) would suggest that we would need a new mechanism of change. If we are unable to find a plausible mechanism, the reason for common descent would remain unexplained, and would perhaps be itself undermined by the lack of a process to explain how it takes place. Natural selection (at the micro level) as the mechanism of change could also be true without common descent being true. In this case, we would then be almost back where we were before Darwin because we would be unable to explain the coming into existence of radically new species. Indeed, those who reject natural selection (at the macro level) are in this position—they usually accept that some type of natural selection can explain changes *within* a species (microevolution), but they deny common descent, and so would need another way to explain how radically new species come into existence.

Most evolutionary biologists believe that natural selection is now well documented, even though in Darwin's time most scientists thought the evidence for selection was weak, and that some other theory of the process of change would eventually be discovered. Darwin thought that natural selection occurred too slowly to be observable,[1] and there are others who think that while it is a *primary* mechanism of evolutionary change, it is not the only mechanism and cannot account for everything about the origin and development of life (especially in the absence of further evidence).[2] In the last chapter, I discussed a few examples of natural selection as a way of explaining the theory, specifically the finch and bacteria examples, and so I will not repeat them here, but just note that evolutionary biologists argue that these types of cases, and many more, have been well documented, and show the process of natural selection in action.

One recent example that is offered as further confirming evidence comes from Peter and Rosemary Grant of Princeton University who studied various species of finches on the Galapagos Islands for a twenty-year period since the 1970s in an attempt to document natural selection in action. They have documented that during changes in seasons, which affect food supply, there are changes in the beak sizes of the finch population. The Grants reason that these changes occur by appeal to Darwin's thesis of natural selection, according to which beak size gives various varieties of finch an advantage at different times of year allowing them to predominate during that time. When environmental conditions change, then the beak size is affected again.[3] Similar conclusions were reached by Michael Singer and Camille Parmesan of the University of Texas, who studied various species of butterfly in Carson City, Nevada; other examples often cited to illustrate natural selection include the change in the body size of guppies exposed to different types of predators, and the way in which flowers are adapted to their insect pollinators.[4] So cases like these are regarded as cumulative evidence for natural selection.

I think it is helpful when discussing the concept of natural selection to introduce a distinction between two kinds of natural selection, or two understandings of natural selection. I have not seen this distinction made in the literature on evolution, but I think it is very useful when thinking about the concept. I will call these two kinds of natural selection, micro-natural selection (MiNS) and macro-natural selection (MaNS). MiNS refers to the process that advocates of the theory of evolution believe takes place when a characteristic in a particular species undergoes change (or "evolves") over time; for example, when the beaks of finches evolve in length over time. Recall that what happens is not that a finch with a short beak develops a longer beak; but rather a species of finches where some have long beaks and some have short beaks to begin with evolves so that the finches with long beaks begin to predominate, due to causal factors involving at least three interrelated things: their prior genetic and morphological makeup, the environment in which they live, and their diets, in the manner explained in Chapter 2. This is one kind of natural selection, and it would account for smaller or micro changes *within* a species; for example, it would explain why the beaks of one particular species of finch became longer over a period of time, or why the body size of guppies changes over time as they interact with their predators. So this kind of natural selection corresponds to microevolution.

MiNS might also explain the coming into being of new species in a simple sense. It could be for some reason (and the "for some reason" is very important because it can be very difficult to identify what the actual causal mechanisms are when it comes to producing evidence to back up these kinds of speculations) that the short-beaked finches and the long-beaked finches were separated roughly into two groups so that no interbreeding took place. We could then say that they are separate species, meaning that they do not breed together, and that one of their distinguishing characteristics is their beak size. But, of course, they are still almost identical in structure (and there is always the question of whether they can still breed together but just don't, say because of environmental factors, highlighting again the difficulty involved in defining what a species is, and when a new one arises).

It could also perhaps be the case that the species of short-beaked finches dies out, leaving only the long-beaked fiches remaining in a certain area. So we might describe these finches as a new species, and distinguish them from the earlier (now extinct) species by their beak sizes, and yet again the species would still be almost identical, and perhaps capable of interbreeding in principle. So MiNS could account in this way for the coming into being of new species.

However, the claims made for the power of the process of natural selection are much more sweeping. In fact, they are incredibly radical. And it seems to me that we can put the more radical claims made about the process in a different category, which I am calling MaNS. The thesis of natural selection makes two further key claims in addition to the fact that specific features of a life form in a species might evolve over time, in the way explained above. The first claim is that this is how *all* features of the organism developed (either directly or indirectly).[5] For example, if we stay with the finches, the heart, eyes, ears, mouth, throat, wings, lungs, feet, toes, nervous, circulatory and digestive systems, right down to the DNA, all evolved in the same way. The process of natural selection, which we explained in Chapter 2, brought about the heart in the same way as it brought about the longer beak. The only difference is the timeline involved; the heart is much more complex than a beak, and so it probably took millions of years more to evolve. The second key point in MaNS is that this process also gave rise to radically different species (macroevolution), not just to smaller changes within a species, or within very closely related species in the way we explained above regarding the finches in MiNS. So MaNS, for instance, leads eventually to a new species descending from the great ape called *Homo sapiens*, and it also gave rise to new species emerging from fish called birds. The process was slow and took many billions of years, and involved thousands of intermediate steps, but nevertheless this is the explanation offered for the existence and development of all species.

I think it is helpful to distinguish between these two types of natural selection for several reasons. The first reason is that they are logically distinct. What I mean is that MiNS could be true, even if MaNS were not true. The second reason is that it seems to me that radically different claims are made with regard to them which justifies putting them in different categories. It seems to be one thing to claim that the features of various species of ape evolved this way but *quite another* to argue that over time the species of ape evolved into man. This is not only because man is a very different species than apes. It is also because MiNS sounds more plausible than MaNS—when I look at the beak of the finch I can follow the argument that explains how the long beak might have evolved (though, importantly, the steps showing explicitly how it happened are not being presented to me, but rather an "explanatory story" with sketchy, piecemeal bits of evidence to back it up). The story of MiNS makes sense to me, and I can compare it to the example Darwin cited with regard to artificial selection used by farmers. But when I apply the *same reasoning* to more complicated features of organisms, such as the eye or heart of the finch, or the human eye or brain, to complex organs and developments in a species, and to the development of one species into another *very different one*, the story told does not sound very convincing. The explanation now being offered does not seem nearly as plausible; it requires a leap to accept the conclusion, and a willingness to be satisfied with vague suggestions and stories instead of actual evidence. Indeed, the argument that this is how the whole life form came into being (and not just complex order, such as in the eye arrangement) seems *incredibly far-fetched*. Many people of goodwill, who reflect on the evidence for this part of the theory of evolution, think that it is not just that it does not seem *as plausible*; it is that it does not seem plausible *at all*, and so requires a lot more evidence to back it up.

Thirdly, all of this brings up more directly the question of the evidence. We can look at both forms of natural selection in terms of the evidence. It seems clear that the evidence for MiNS is better than for MaNS; there is some evidence for MiNS but there is very little for MaNS. The evidence for MiNS that is often cited includes familiar examples that are common in biology textbooks and that we have already mentioned, such as the beaks of finches, the evolution of fruit flies, mice, and bacteria. Yet the evidence overall for MiNS is still sketchy. In making this point, it is important to keep in mind the difference between *explaining* how MiNS is supposed to work and giving *direct evidence* that it did in fact occur this way. It seems to me that there is not much actual evidence to support even MiNS, though there is some. But it is worrying that there is little direct evidence of the sort that people would usually need to be fairly convinced. The examples offered to support MiNS usually lack specifics and require a lot of extrapolation.

Some evolutionary biologists deal with the difficulties facing MaNS by claiming that there is really no difference between MiNS and MaNS, that MaNS is just MiNS on a larger scale. While there is some truth to this claim, this reply does not address the problem of the implausibility of extrapolating from MiNS to MaNS without further evidence.[6] It is not long before we start running into speculation about how it *could* have happened as a substitute for actual evidence showing *how* it did happen. One can see this when one examines even the supposedly simple example of the beak of the finch and how difficult it is to document even this small claim about MiNS. There is little actual empirical evidence involved. And how could there be since most of the fossils and other essential evidence we would need to know how finches that survived, developed, and reproduced is long gone, including evidence relating to the environmental habitats, climate patterns, food types, competitors of the finches, not to mention detailed evidence about the biological structure of the finches' ancestors. *Even if this evidence were available*, it would still be difficult (indeed impossible) to reconstruct exactly the path of development of the finches. The story about the finches is based on what lawyers would call circumstantial evidence. We have fossils of finches with different size beaks, we know something about the climate in which they lived, about the food available to them (both now and in the past), about their ancestors, about their competitors. But we still have to reconstruct much of what we think might have happened in their evolutionary development not only to explain what the theory of evolution is saying, but also to show that the theory is true (and once again we see the confusion between these two very different points). Evidence for the second point would support the theory of natural selection, but telling a story about the first would not.

Darwin did acknowledge that his argument for natural selection would face problems and would likely appear implausible to many, not only in terms of the evidence, but even conceptually. He put it like this, in a famously quoted passage: "To suppose that the eye, with all its inimitable contrivances for adjusting the focus to different distances, for admitting different amounts of light, and for the correction of spherical and chromatic aberration, could have been formed by natural selection, seems, I freely confess, absurd in the highest possible degree." However, he thought these difficulties could be surmounted because we can find in nature numerous cases where one function of an organ was later replaced by another function in its descendants: "If it could be demonstrated that any complex organ existed, which could not possibly have been formed by numerous successive, slight modifications, my theory would absolutely break down. But I can find no such case .... Numerous cases could be given amongst the lower animals of the same organ performing ... wholly distinct functions."[7] He discussed several (mostly speculative) examples involving barnacles, insects, and the swim bladder in fishes that was originally used for flotation, but then later for respiration (and so he thought natural selection could further convert it into a lung in higher animals).

The question is: could it? Or to ask the question more precisely perhaps: is it reasonable to believe that it could? Most evolutionary biologists today answer this question in the affirmative. Dawkins, for instance, says that, "... not a single case is known to me of a complex organ that could not have been formed by numerous successive slight modifications."[8] However, a problem arises because we do not know in most cases what advantage each evolutionary development conferred on a species, and are forced to speculate by introducing a lot of what critics have described as "just so" stories to explain how natural selection works in various species. So we would speculate then about what role the various parts of the eye were performing before they came together to collectively produce the function of sight, but in most cases we do not know *the actual role* they were performing or indeed if they were performing *any* role; nor do we know the successive steps in their pathway of evolution. This leaves a gap in our account of natural selection, and it seems to be a gap that is very difficult, perhaps impossible, to fill. This is why the evidence for natural selection is more like various parts being brought together to build a cumulative case; rather than evidence being available and transparently clear that demonstrates the conclusion and leaves little room for doubt.

So I think it is misleading for any scientist to say that they are "certain" that evolution occurred, or that they don't "believe" in evolution, because in fact they "know" it is true, and so forth. This can only reflect their own commitment to the certainty of the theory, but it is not an accurate report on the state of the evidence. Many books discussing evolution are prone to exaggerate and overstate the extent of the evidence, and to either downplay or, more usually, to ignore problems, especially with regard to the question of how natural selection could produce complex organs, and concerning the absence of fossils of transitional species. The quote from Dawkins above is a clear example of this, since in fact he has no cases—*none whatsoever!*—where he can explain how natural selection brought about the evolution of complex species, or of complex features in an organism. It is no wonder that he goes on immediately after the above remark to *invent* speculative stories about two cases (concerning fish, and the eye) to illustrate his claim! These stories are based on Dawkins's *assumption* that his understanding of natural selection is true, but they are not a *demonstration* of natural selection. A difficult problem for the case of natural selection is that it seems to be quite impossible to reconstruct the *actual* processes of evolutionary change in any particular species at the macro level. This is why Dawkins and others are prone to invent examples to illustrate natural selection at the macro level, and then to gloss over the fact that these examples are not real, but fictional, cases.[9] Ernst Mayr claims that "... it has been shown that natural selection is capable of producing all the adaptations that were formerly attributed to orthogenesis [and other teleological explanations]."[10] Of course, nothing of the sort has been shown, since he is relying on extrapolations and imaginary stories, not evidence from real cases. But until one can present specific cases, there must be doubt over the general extrapolation argument. And if one were to reply that, in fact, we can *never* reconstruct the specific explanations, then it seems that we have a general problem in the theory.

To defend what I have been calling MaNS, evolutionary biologists often cite the example of the development of resistance to antibiotics in bacteria, the example we described in Chapter 2. This example is much championed as evidence for MaNS (not MiNS) in expositions of evolution. The reason for this is because it is so difficult to find examples of macroevolution, for obvious reasons. The long timelines involved in the process, and the difficulty of obtaining the other knowledge required, such as of geological effects, environmental conditions, food supply, prey, and predators, all make it hard to document actual cases of macroevolution in action. And there is something logically unsatisfying about saying that it is true but because of the nature of the process, the direct evidence no longer exists. (It is for this reason that Darwin's arriving at the theory of natural selection, and his attempt to argue for it, must be seen as an incredibly intelligent and imaginative piece of reasoning, whether the theory is true or not!) So in an attempt to work through an actual case of evolution, biologists will often turn to the case of antibiotic resistance in bacteria because it is claimed that the processes involved give rise to new species of bacteria, not just to changes within the same species. The reason it can give rise to new species is because bacteria exist in huge numbers and multiply rapidly, and so they are useful to study from an evolutionary standpoint. But, as we noted in Chapter 2, the bacteria in this example do *not* develop resistance as a reaction to the drug; rather, there was a species of bacteria already resistant to the drug. These bacteria then came to dominate, as the drug eliminated the ones that were not resistant. Also, sometimes when a patient takes a drug, none of the bacteria that are present in his body may be resistant but because of the high level of reproduction and mutation, bacteria eventually appear that are resistant. This has nothing to do with a reaction to the drug; bacteria come along due to the strains of original bacteria, the rapid rate of reproduction, and all the other variables involved in the causal process.[11] This case is often offered as evidence for natural selection and for macroevolution; for the former because it shows how a particular species dominates over others, and for the latter because it leads to new species of bacteria.

The bacteria example, while good evidence of natural selection in the sense of MiNS is not good evidence of MaNS, or therefore of macroevolution. It is true that the two concepts are very closely related, at least in the way that evolution is currently expressed and developed. But as we noted above, although the two are closely related, they are logically distinct because it is logically possible that microevolution could be true (evolutionary changes within a species), but macroevolution false (evolutionary changes leading to radically new species), and if so, some other mechanism would be needed to explain how radically new species come into existence. But as the theory is currently expressed, it is argued that if one has a mechanism for explaining how microevolution occurred, some at least think this makes macroevolution more believable. It seems as if for some, especially in biology, MaNS seems very plausible given MiNS. This is the conclusion of Nobel Prize winning biochemist, Christian De Duve. After describing in some detail the process of how bacteria mutate and adapt, he concludes that "Bacteria can behave this way because of their rapid proliferation rate. Note, however, that the bacterial strategy *has also been followed by more complex organisms, but much more slowly and gradually, over eons of time; variation screened by selection is the mainspring of Darwinian evolution*."[12] Note De Duve's extrapolation from what we can conclude from a study of bacteria to the conclusion that the same process takes place on a much larger scale. (Note also that he does not present it as a reasonable claim, or a plausible explanation, but as a fact.) The question is: is it correct to read the bacteria example as evidence of macroevolution, rather than just microevolution? This appears problematic to many because although one gets new species of bacteria through natural selection in this example, the species are not radically different enough to count as MaNS in my understanding of the concept. The new species are still bacteria, and indeed are essentially the same in general biochemical structure as their ancestors. We still have only bacteria after the process of rapid multiplication has taken place, and not flies or insects or fish or birds or dogs. So while one can grant that this is an interesting example of MiNS, it is not an example of MaNS, and, as indicated earlier, it is not clear whether we have any clear and undisputed examples of MaNS.

So it is fair to say that there are gaps in the evidence for natural selection in several important places. And because the claims for natural selection are so radical, these gaps worry many who wish they had stronger evidence in this area of the theory. One question to think about is whether the evidence for microevolutionary processes that lead to evolutionary change within a species is good evidence that macroevolutionary change occurs across widely differing species? *Is the extrapolation from the former cases to the reasonability of believing in the latter cases justified*? Many scientists argue that it is justified, especially when coupled with all of the other evidence we have. When we look at all of the evidence collectively, biologists will argue, even though one or two areas, such as that concerning macroevolution, might be weaker than others, it is still reasonable to conclude that the main claims of the theory overall are reasonable to believe. On the other hand, many look at some aspects of the theory and, it must be said, are not quite convinced. And this is not just because they are motivated by religious concerns, or are ignorant of the evidence; it is because they don't think the evidence is strong enough to warrant the claims made based on it. Perhaps we can say that within biology there is a strong consensus that the evidence for macroevolution is strong, but that outside biology (and perhaps even in other parts of science), there is not quite so much of a consensus. Many are also not that impressed by the fact that there is a consensus of scientists who think the extrapolation is justified, which is unusual because normally outsiders would be closely guided by the views of the scientific community. There are two reasons for why the consensus of scientists does not convince people with regard to macroevolution: one is the fact that the theory of evolution comes (as we have seen) with many worldview, ideological and political implications, and many think that these implications play a role in how leading evolutionary biologists read the evidence (this is especially true with regard to whether biology shows evidence of teleology); the second is that there is something of a climate of fear around the theory today that may prevent a dispassionate discussion of the evidence. If it is the case that one fears negative consequences, or even losing one's job as a science teacher, for voicing criticisms of evolution, then it is obvious that one will hear few criticisms of it.

We must also raise here, but will not discuss until later chapters, questions about how natural selection can account for certain features of species, especially the human species. It is one thing to suggest that it might explain the biological features of species, but what about advanced features such as those *in our own species*: for example, human consciousness, including our capacity for reason, free will and morality, and related features? These features of the human species present special problems for evolutionary explanations, as we will see later on, and they add to the worries many have that natural selection cannot be the whole story about human evolution.

What is the evidence for common descent?

In the Jesse James Museum in Kearney, Missouri, there is a diagram of Jesse James's family tree. It is fairly detailed, showing Jesse and his ancestors near the top end of the diagram and his descendants at the bottom. It documents his parents as his immediate ancestors, and then their parents on both sides (his grandparents), and the ancestors of his grandparents going back several generations. It also traces his siblings (including his brother Frank), and all the children, including Jesse's, and their children, down to recent times. This is a diagram of the tree of life of the extended James family before and after the birth of Jesse. All of the currently living members of the James family are therefore genetically related to every preceding member of the James family by common descent. One can trace a line from a currently living member of the family genetically right back to the first couples listed at the top of the diagram. Suppose we were not sure if a currently living person belonged in the James family tree, how could we find out? Even if we had access to them, skeletal fossils of James family members probably would not help us since they would be too similar in structure to all human skeletons; but DNA and other comparative chemical tests on both the fossils and the currently living person would provide us with some information on which to base a judgment. Darwin as we know argued that all living things were united by common descent, just as with the James family. All current life forms and species are related genetically to earlier life forms and species, even to bacteria and plants, going back at least 3.5 billion years to when it is thought that the first living things appeared on earth. So our question is: what is the evidence for this conclusion, particularly the claim that all animals are related to each other by common descent?

There are two main lines of support for the thesis of common descent (or descent with modification, or common ancestry), along with several other bits of evidence, all of which, evolutionary biologists claim, reinforce each other with cumulative effect. The first important source of evidence is the fossil record. Recall that fossils are the remains of organisms, life forms, and species that have been preserved in rocks, plants, swamps, and other parts of nature. Fossils come in different forms; the main categories include the "hard" parts, such as skulls, bones, and other skeletal elements that are preserved either in whole or (more usually) in bits and pieces, and the "soft" parts—the cellular tissues of an organism—that are sometimes preserved with the hard parts. Fossil traces are also explored, such as imprints, molds, and footprints, and are a good source of supplementary confirmation. Excavating and classifying fossils is obviously a laborious process, but an immensely important one for paleontologists. Examples of fossils of various species that have been discovered would include fossils of horses, dogs, whales, chimpanzees, elephants, and man.[13]

How does a study of fossils show us (i) that similar species within the same genus are likely related, and (ii) that widely different species are related to each other by means of common descent? The argument is a fairly straightforward one. The first point paleontologists will make is that the fossils of different species may have similar anatomical structures (for example, modern elephants, 10,000 years old, and extinct elephants, 34 million years old, have very similar skull and trunk structures; a similar argument is advanced for the evolution of the horse, and other species). It is reasonable to conclude from this fact alone, the argument goes, that these species likely have a common ancestor, that the later species are descendants of the earlier ones, and that this is why the younger ones have very similar traits to the older ones. This argument is often stronger than it appears at first sight because paleontologists have stressed that we are not just looking at fossils in general outline, where similar structures and appearances might not necessarily indicate that one species was a descendant of the other. What we in fact find with many fossils, it is claimed, are quite distinctive features that appear to have been passed on in successive generations (for example, the development of the mouth and jaw structure over a 3–4 million year period that eventually appears in modern humans). The second thing to consider is the timeline of the fossils. Dating of fossils is done by means of measuring the amount of radioactive decay present in the elements in the fossil, such as carbon, though the measuring techniques are not without problems.[14] This will then help to place the fossils in sequence from oldest to youngest so that one can check for anatomical similarities and development. The third point is that the geography must be right. It must be geographically possible for one species to be a descendant of another (the fossils are found in the same general area, they indicate similar habitats, diets, climates; for example, the lineage of the armadillo).[15] When these three points are established with regard to various species, it is reasonable to conclude that the fossil record shows that some species are descendants of others (and that this is more reasonable than saying that each species in a documented lineage arose from what Darwin called acts of "special creation"[16]).

This line of reasoning can be extended to the lineage of human beings. At the time of Darwin, there were no fossils linking human beings with apes, but since then, some fossils have been found, and the closer to the present time these fossils are, the similar they are to modern humans. This lineage includes *A.afarensis* (4 million years ago), *A.africanus* (2.5 million years ago), *H.egaster* (1.5 million years ago), *H.heidelbergensis* (1 million years ago), and *H.neanderthalensis* (500,000–250,000 years ago). A similar inference is then drawn: that these species are recent ancestors of human beings, and that we are their descendants by genetic inheritance.

Is the DNA evidence for macroevolution persuasive?

Since the 1950s pioneering work in genetics and molecular biology has led to the view that each species has a set of biological instructions that are contained in the molecules and cells of each organism. These molecules are called deoxyribonucleic acid, or DNA. DNA was discovered in 1871 by a German biochemist, Friedrich Miescher. In 1953, four scientists, Francis Crick, Rosalind Franklin, James Watson, and Maurice Wilkins, discovered the double helix structure of DNA—a twisting ladder-like sequence that contains the biological information, or "genetic codes," that are passed from one generation to the next in sexual reproduction.

DNA is found in molecules inside a special area of the cell called the nucleus; the DNA molecule is tightly packaged, and is known as a chromosome.[17] Chromosomes, which are really bundles of DNA, contain some strands of DNA material called genes; the genes contain instructions to make proteins, which are the building blocks of life. Chromosomes come in pairs in an organism, one half of each pair contributed by each parent, so an organism inherits half of its DNA from the male parent and half from the female parent. The complete set of DNA in an organism is called its genome (which includes all of its genes). The actual chemical building blocks of DNA are called nucleotides, which themselves consist of three parts: a phosphate group, a sugar group, and one of four types of nitrogen bases.

The gene carries the vital information necessary for the synthesis of proteins necessary for the structure and function of organisms. Proteins determine or heavily influence many features of the biological and behavioral characteristics of an organism, including its physical appearance, how well its body metabolizes food or fights infection, even how it behaves. The size of an individual gene may vary greatly, ranging from about 1,000 to 1 million bases in humans. The complete human genome contains about 3 billion bases (denoted by sequences of letters), which includes about 20,000 genes in twenty-three pairs of chromosomes (the DNA in each pair in the form of a double helix). A mouse genome has 3 billion bases, and a worm has about 100 million. The discovery of DNA has been very important for the theory of evolution. The existence of genes in various species explains the passing on by parents to offspring of traits that later turn out to be advantageous for survival. Alterations in the genes of an individual—which are brought about by the process of mutation—would affect the trait that has been passed on, and therefore may have an effect on the survival of the organism. Mutations, as we noted in Chapter 2, are caused by problems inherent in the DNA itself, as well as by environmental factors, including diet, climate, and so forth.

It is argued that the makeup of the gene through an analysis of its DNA provides further evidential support for evolution. Recent work in genetics shows that there is 60–70 percent similarity between the DNA of human beings and mice, approximately a 94 percent similarity between the DNA of human beings and chimps, and an even closer similarity between the DNA of

Neanderthals and human beings. This suggests, therefore, that such species are genetically related, based on the same principle that enables us to tell that two men are brothers, or that one man is the father of another, the basic idea being that the closer the genetic similarities are the more closely related the species. This is the same sort of analysis that would tell us that a currently living person is related to Jesse James, providing we had access to the DNA of Jesse (which we do from the exhumation of his remains in Kearney in 1995!). We can now document by an analysis of DNA that part of the tree of life that illustrates the lineage of human beings, chimpanzees, gorillas, and orangutans, all the way back to the common ape-like ancestor of all these species. Combining the fossil evidence, the evidence from comparative anatomy, along with DNA evidence, and other evidence, is called "the modern synthesis."[18] The modern synthesis means, among other things, that the fossil evidence and the DNA evidence cohere; for example, given our reconstruction from the fossil record of the family tree of apes, we find that the chimpanzee is more closely related to the bonobo than to the gorilla, so we should find the same result when one examines the DNA of all three; the DNA of the bonobo should be more similar to that of the chimpanzee than to that of the gorilla; and this is what the DNA analysis does show, according to the latest analysis.

One might wonder if the DNA evidence actually shows what it seems to show. Critics might argue that God could have designed all life with a similar DNA structure, and so it would not follow logically that species with similar DNA are genetically related to each other. The DNA evidence shows that various biological structures are even more similar than we thought, but does it follow that they are genetically linked? As an analogy, we observe great similarities between new models of the same make of car each year (usually moving from the simple to the complex), but we don't infer that they are genetically related; we know that the reason they are similar is because they have been designed that way! Might it not be the same with nature?

Francis Collins, former head of the Human Genome Project, acknowledges this point, but he argues that the similarity in DNA is so striking that it inevitably points in the direction of common descent. Collins compares the mouse genome with the human genome, both of which have been determined to a high degree of accuracy. Several features point to common ancestry, Collins argues, including the similar size of the genomes, the remarkable similarity of the inventory of protein-coding genes, and the ordering of the genes in both, which is substantially the same over long stretches of DNA (e.g., all of the genes on human chromosome 17 are found on mouse chromosome 11). There are also similar sequences of "junk DNA" in both mouse and human, and the clincher for Collins is that sometimes this

DNA is damaged (and as a result cannot function), and "in many instances, one can identify a decapitated and utterly defunct section of DNA in parallel positions in the human and the mouse genomes."[19] Collins recognizes that some will object that "junk DNA" may not really be junk at all, and that the ordering of genes in the chromosomes is crucial for development and so the different ordering in different species should not be taken as a similarity but as a difference. Nevertheless, he thinks that "unless one is willing to take the position that God has placed [this DNA] in these precise positions to confuse and mislead us, the conclusion of a common ancestor for humans and mice is virtually inescapable."[20] Jerry Coyne observes that "Every fossil that we find, every DNA molecule that we sequence, every organ system that we dissect supports the idea that species evolved from common ancestors."[21]

There are other bits of evidence that can also be used to provide further support for common descent. These would include: (i) the similarity between the anatomies of different species (e.g., between the forelimbs of whales, birds, dogs, and human beings); (ii) the more general similarities between species (e.g., in their digestive, circulatory, and nervous systems); (iii) the fact that the embryos of different species have some similarities in their early stages (e.g., human and dog embryos); (iv) the fact that the fossils over time become gradually more complex; and (v) it may account for the presence of vestigial (or redundant) organs in some species (the appendix in man, or the wings of ostriches).

These last bits of evidence are crucial to the overall case for evolution because they provide some support for macroevolution on a large scale, e.g., from fish to birds, and from plants to animals. This "synthesis" of evidence can be appealed to indirectly to support macroevolution in the sense that it would show evidence of common descent across very different species, but it would not by itself be evidence of natural selection as the mechanism of species change.[22]

A problem for evolution?

The problem of the lack of fossils of what are called "transitional species" has generated much discussion with regard to the question of the evidence for evolution. Many thinkers, including Ernst Mayr, Jerry Coyne, Stephen J. Gould, Peter van Inwagen, and Samir Okasha, have considered the question.[23] A transitional species is a species that might be described as the common ancestor of its descendants; the descendant species would differ in significant ways from the original species because they would have quite different traits. The ancestor would have some mixture of the traits from *its own ancestors* and of the traits of its descendants. Although discussions of the evidence for evolution are often vague on this matter, I think we can distinguish between two broad categories of transitional species. One would be those transitions within species that have similar general traits, such as within the various species of whales. Modern whales are descended, it is claimed, from earlier whale-like creatures over at least a 50 million year period. This means that in addition to fossils of well-documented recognizable species of whales, there should be fossils of those transitions that led to these species (indeed, in principle, there would be millions of such fossils between the start of the process that eventually led to the emergence of modern whales at its conclusion). The second type of transitional species would be those that existed (according to the theory) between species that are more radically different from each other (macroevolution), such as between reptiles and birds, or between apes and human beings. Again, there existed, if the thesis of common ancestry is correct, millions of such fossils. Given that many of these types of fossils must have existed in the past, one would expect to have a rich repository of evidence of transitional species to study. Yet it is the case that there are few indisputable and uncontroversial candidates. Of course, many of these transitional fossils will have been lost for a variety of reasons but nevertheless it seems to be a very reasonable conclusion that the absence of numerous and a wide variety of transitional fossils is clearly an anomaly in the theory of evolution. Their absence is odd since we do have plenty of fossils of the non-transitional sort, fossils of our more recognizable animal forms and groups. Surely if we have an abundance of these, we should have an abundance of the transitional fossils too?

Darwin did acknowledge that the absence of transitional fossils was one of the main objections to his theory. As he put it, "But, as by this theory innumerable transitional forms must have existed, why do we not find them embedded in countless numbers in the crust of the earth?"[24] His answer that there would not in fact be all that many transitional species, and that they would likely be confined to narrow geographical areas, is far from convincing. Contemporary evolutionary biologists will sometimes acknowledge this problem, but often downplay it, as Gould has noted.[25] Their usual reply is to insist that they have discovered hundreds of transitional species. Two of the best known are *Archaeopteryx* and *Ambulocetus*. *Archaeopteryx* is a fossil of the earliest bird, which lived around 150 million years ago (in Germany), and is evidence of a transition between dinosaurs and birds. *Ambulocetus* (dated to be 50 million years old and discovered in Pakistan) is a whale-like crocodile, which seems to have been amphibious, and is thought to be a common ancestor of modern whales and land mammals. Biologists also sometimes suggest that earlier species in particular, which may have consisted predominantly of soft parts, were unlikely to fossilize. These are interesting points no doubt but they still leave us with large gaps in the fossil record for which we have no convincing explanation; and even though some early species would not have fossilized, this point would not help us with the ancestors of whales or elephants, for example, which would have fossilized.

There is also the problem that the Cambrian explosion raises for the theory of evolution, which is directly related to the problem of transitional species. Around 700 million years ago, as we noted in Chapter 2, we find the *rapid appearance* of many major species ancestral to several modern species, and by the end of the Cambrian period, there is evidence for nearly all major animal groups. The Cambrian period appears to provide evidence against MaNS because it suggests that the major animal groups did not evolve gradually over millions of years but appeared suddenly as it were in the fossil record. The absence of fossils of the species that were ancestors to the species found in the Cambrian period has often put evolutionary biologists somewhat on the defensive. Darwin admitted that the problem of transitional species confounded him, and it still worries many today who are otherwise favorably disposed toward the theory.

There is also a more general question about the evidence for macroevolution that transitional species may not help us with. For example, if the tree of life as it is represented in most biology textbooks is accurate, this would mean that life started as a single-celled organism, which then divided and reproduced, and eventually we ended up with all of the species we now have (including us), as well as the millions that became extinct along the way (90 percent of all species, according to some estimates). This is a very speculative claim that would require many detailed transitional fossils in order to lay it out clearly, and indisputably. This evidence would also have to show how the two sexes originated, and how they became perfectly reproductively compatible. Biologists argue that the fossil record will eventually establish these claims as it becomes more complete, and critics argue that until we do have this evidence before us, macroevolution must be regarded as more speculation than fact.

Is evolution only a theory?

One often hears the charge that evolution is "only a theory." This charge also reminds us of our discussion in Chapter 2 concerning the distinction between scientific theories and facts, and about objective knowledge and proof in general in the discipline of science. We need to ask if evolution is a theory, whether it is "only a theory," and whether this means that it is not a "fact," and might be disproved later. For the charge that evolution is only a theory is often meant to imply that it has not been proved, and so we cannot say with confidence as many evolutionary biologists do that it is true; we are therefore entitled rationally to reject it, and even propose rival theories. Rival theories would then have to compete with each other in the discipline of science, and within the wider culture; the theory of evolution would not then be able to claim the high ground because it is "only a theory" like the others, despite the protestations of its proponents to the contrary, and despite the occasional supercilious attitudes of those who talk about evolution as if it is a "fact."

However, the claim that evolution is "only a theory" is too simplistic and is not the whole story about theories, facts, and proof. To appreciate this point, we need to make a distinction between well-confirmed theories and poorly confirmed theories. We also need to recall our distinction in Chapter 2 between facts and theories. We noted in our previous chapter that there are two approaches to theories. The first is that the facts and the evidence one has gathered may *suggest* a theory to explain them. For example, Galileo's observations of the movement of the planets, along with his mathematical analysis, suggested to him that the Copernican theory of planetary motion was correct (in short, suggested to him that the earth itself was moving). The other approach is that one proposes a theory to explain some phenomena, and then one seeks facts and evidence to support the theory (the Higgs boson theory in particle physics would fit into this category). Most scientific theories adopt elements of both approaches: that is to say, the facts and evidence may suggest a theory, then the theory helps to further explain the facts, to interpret the facts and evidence in a new light, and perhaps to suggest further confirming evidence to look for, and so forth.

The appraisal of a theory must be done logically and rationally. First, this means that one must look to see if the facts and evidence do support the theory, and if the theory does explain the facts well, taking into account many disparate things (such as explanatory gaps, anomalies in the evidence, and how the theory fits with other well-established theories). Second, one's assessment of these matters should be free from biases, financial considerations, ambition, or ideology. Third, this is also why it is important that there be a communal role to the rational appraisal of theories, because it helps to safeguard against these temptations, against mistakes or prejudices that might motivate one individual, or a group of individuals, to interpret the evidence in a biased manner. A theory that satisfies these requirements can be described as a well-confirmed theory. This means that the evidence does support the theory quite well, that there is a rational, logical fruitful interplay between facts and theory and between theory and facts, and that a consensus of the best minds who work on the theory agree that it is a well-confirmed theory. A poorly confirmed theory, on the other hand, would be one which fails on these three points; an example would be theories concerning astrology.

Now, looking at evolution in terms of this understanding of theory, we can make three important points. First, the evidence suggests the theory of evolution. The evidence consists of the material we have just reviewed: fossil evidence, DNA evidence, other bits of circumstantial evidence, and the synthesis of independent lines of evidence. Darwin and his successors looked at this evidence and it suggested to them common descent (at least), but also natural selection. Second, the theory gives *us a new way to look at the evidence we have*; it brings together lots of different bits of evidence and gives them a coherence and overall explanation. For example, if we assumed common descent, we could explain why some species look very similar in structure, predict that they would likely live in the same location, and so forth. So if we found a fossil of an older elephant skull with certain features, we could predict that we might find a fossil of a younger elephant skull with similar features before we even begin to search for it. In this way, the facts suggest the theory, and the theory explains the facts. The third point is that there is a consensus among scientists who work in the area of evolution that the facts and evidence do support the theory.

So we can conclude that evolution is a well-confirmed theory. This means that the facts and evidence do support the theory of evolution, and that the theory also helps to explain the facts and evidence fairly well. It also means that many people who have worked closely on the theory agree that the evidence for it is strong. It does not mean that the theory is proved; that there can be no doubt; that there are no explanatory gaps, problems, or that there is nothing further to be explained (despite the way evolutionary biologists sometimes talk). Nor does it follow that there is no contrary evidence. Often there is contrary evidence to a theory (often called anomalies), but a theory can tolerate a small amount of contrary evidence if the positive evidence is strong enough. Supporters of the theory then claim that they will eventually discover how to account for the anomalies within the theory. An example of an anomaly within the theory of evolution would be the lack of transitional species, discussed above, with supporters claiming that, despite this, the evidence for macroevolution is very strong, and that we will eventually discover the reason for the lack of transitional species (or a scientist may propose an explanation for why there is a lack of transitional species, e.g., Gould's theory of "punctuated equilibrium"). But we should note that in the review of evidence for evolution, there is sometimes a tendency among evolutionary biologists to exaggerate the number of transitional species, or to downplay this problem, in the face of criticism (as we noted in the previous section), rather than admit to it, acknowledge that it is an anomaly that has to be faced. This is an example of the biases some may bring to the theory preventing them from looking at the case objectively. And the occasional reluctance of evolutionary biologists to discuss the problems for evolutionary theory, in addition to presenting the positive evidence, worries many, and only raises further questions. (It is rare, for example, in discussions of evolution in books of popular biology, including textbooks, to find a sustained, honest discussion identifying the problems with the theory, in the way one will find with regard to the theories presented in physics textbooks. On the contrary, the evidence is presented often *as if there are no major problems, and no room for doubt*; difficulties are papered over and lightly dismissed. Perhaps biology textbooks should include a section listing and discussing the problems and issues of contention in evolutionary theory?)

This brings us to the point we have been leading up to. We can now answer the question posed in the title of this section by saying that it is not accurate to describe evolution as "only a theory" for the simple reason that this implies it is a poorly confirmed theory. But it is more accurate to describe it as a well-confirmed theory. And this is all it needs to be taken seriously, to be given respect, and to be regarded as true (with the understanding that it is not "proved"). We cannot therefore attempt to put evolution down by claiming that it is "only a theory." It is true, of course, that evolution has not been proved, but, as we noted in Chapter 2, it is very difficult to claim that any scientific theory is "proved." The best we can hope for is to say that a theory is well confirmed, and also perhaps that it is the best theory we currently have to explain the facts. Several leading biologists believe that the theory of evolution is so well confirmed that we can call it a theory and a fact![26] But a fact is normally understood to refer to something that is truly the case, something that is verifiable or proved and cannot be doubted. So it is an exaggeration to describe evolution as a fact, since it has not been proved to this extent, and as we just noted, it seems to be impossible to prove that any theory is true in science in this way (a point that many scientists usually acknowledge in the abstract but then reject when they start talking about their own pet theories!).

But it is important to conclude this section by noting that this kind of exaggeration by various biologists is irresponsible, and actually in a paradoxical way serves as fodder for those who would claim that evolution is only a theory. Skeptics can effectively reply by pointing out that the theory is not proved, going on then to dispute what evidence there is for it, ignoring a distinction that leading biologists too should have emphasized: that even though the theory is not proved, the evidence for it is very good. In their zeal to promote the theory of evolution, many biologists ignore this important distinction. I think we have to acknowledge that in current discussions there is a regrettable tendency to exaggerate the evidence in favor of the theory of evolution; this practice makes it difficult to have a clear discussion of the evidentiary question, and it also adds to our cultural problems with regard to the theory. Unfortunately, many evolutionary biologists regard as their main critics in discussing evolution, not other scientists, but creationists, and this tends to make them defensive, prone to exaggeration, and to downplay or paper over problems, or present them as trivial. Perhaps this reaction is understandable from a human point of view, but from a logical point of view it only makes an honest appraisal of the theory more difficult.

Can evolution be falsified?

We turn now to an interesting charge that has often been made against the theory of evolution, a charge that deserves a separate discussion. This is the claim that evolution is a theory that cannot be *falsified*. In order to understand the point of this charge, we need to take a minute to explain the notion of falsification in science. The idea was first proposed by philosopher of science, Karl Popper, who was interested in the question of how to judge whether a scientific theory is a good theory or not. Popper was interested in thinking about a list of criteria that a theory would have to satisfy to be judged a good theory. The criteria he had in mind were common sense ones like evidential support, testability, ability to make predictions, coherence with other theories, and so forth, but he also proposed an important logical criterion: falsifiability.[27] This criterion was based on a logical claim that Popper thought applied especially to scientific theories. A scientific theory claims that certain facts are true of the world, or are very probably true, so this means logically then that certain other factual claims cannot be true. So, for example, if Newton's claim that for every action there is an equal and opposite reaction is true, then it would be false that we could have an action in the world, all other things being equal, that could produce double the reaction, or if it is true that water boils at 212°F, then it would be false that it would be possible for us ever to find a case of water boiling at 80°F.

Popper made this point by saying that if we assert that a particular claim or theory is true in science, say the claim about boiling water, then we should also be able to say which facts would, *if discovered*, falsify the claim or theory. So Popper called this the criterion of falsifiability, and argued that for any scientific theory, we should be able to state, in order for it to be a good theory, what evidence, if discovered, would falsify the theory. Two other conclusions follow as well: one is that if we did not find the opposing evidence over a period of time, this would count strongly in favor of the theory; the second is that if we could not state *in principle* what would count as falsifying evidence, then it would mean that our theory was suspect. This is because nothing would appear to be able to falsify it, and this is equivalent to saying that any piece of evidence one finds can actually satisfy or confirm the theory. Popper thought this conclusion was illogical with regard to scientific theories. As examples of theories that failed the criterion of falsifiability test, he listed psychoanalysis and Marxism. In the case of psychoanalysis, he argued that any piece of evidence that one presented against the theory could be turned into evidence to support it. So we are unable to state, even in principle, falsifiability conditions for the theory of psychoanalysis.

Popper also raised a question about whether the theory of evolution might be a theory that failed the criterion of falsifiability, especially the thesis of natural selection. Initially, he thought it might, but later was not so sure.[28] The criticism of evolution is that it is very difficult to state falsifiability conditions for the theory. Indeed, it looks like any kind of evidence that one finds in nature could be interpreted to support evolution, or pressed into service to support evolution, or at least read as not counting as evidence against it. If this were true for every bit of evidence that one considers, this would mean that logically the theory is suspect because it cannot be falsified. Let us look briefly at some hypothetical examples as possible counterevidence to evolution. First, suppose we found a fossil of a species we had never seen before, or a fossil that had an unusual feature that its supposed ancestor did not have, or a fossil with an unknown and unusual structure, would any of this evidence go against the theory? It is easy to see that nothing along these lines would count as contrary evidence because findings of this sort are all compatible with the theory as currently stated. They could be explained as simply discoveries of new species, or of new lineages from previous descendants that had not yet been classified, or as cases of evolutionary development that had taken a radically different turn (for some unknown reason) somewhere in the development of a particular lineage. So no evidence of this sort could be used to falsify the thesis of common descent. As another possibility, suppose we thought that the ancestors of horses had required a certain type of climate to survive, but later evidence points to the fact that they did not have this climate. Or suppose we found that our understanding of the evolutionary lineage of *Homo sapiens* was wrong because of new fossil discoveries, would any of these issues raise problems for the theses of common descent or natural selection? It is clear that none of these objections can work because biologists will simply claim that we just don't know enough about the environment of previous eras, or about how species interacted with their environments, or that our taxonomy was incorrect, or that the evidence is still incomplete (which of course it is, and will be permanently).

Let us look at a third possibility. Suppose we discover a fossil of a dog that had two stomachs and that appeared to be—in terms of timeline and geography—in all other respects an ancestor of modern dogs, would this count against the theory? Or would many examples of this type count against the theory? It seems not. Would not biologists simply say that evolution must have taken a radical turn, which it often does (isn't this what the Cambrian explosion shows?), and that unusual environmental factors, including diet, had a causal influence on DNA structure to produce these kinds of radical change? Of course, we could not say precisely how such changes came about because the detailed evidence is long gone (as it is with regard to most claims of the theory). Again, it seems as if even radical difficulties like these could be incorporated into the theory, even though there would not be much hope of *ever* being able to explain them. Fourth, the biologist J.B.H. Haldane is supposed to have said that the discovery of a rabbit fossil of the Precambrian period would destroy his belief in evolution. But is this really a case of a bit of evidence that would falsify evolution? Would it not mean that we would only have to rethink out taxonomy of various species, to reclassify our timeline? Haldane's point is that this fossil would be so out of sequence that we would have to question the whole theory. But would we question it? I think we would all agree that if we actually found a species of rabbit in the pre-Cambrian period, it would be very unlikely indeed that scientists would question evolution; they would simply say that their systems of classification are incomplete and perhaps inaccurate, that the dating system may be wrong, and so forth. In short, they would do anything other than question the theory!

But let us consider a fifth case. Suppose we found a very complex fossil, indicating a life form as complex as a chimp or a human being, that dated to 4 billion years ago. This would be a good candidate for falsifying the theory. This is because the theory claims that life began about 3.5–4 billion years ago, and that it moved from simple to more complex life forms over the next several billion years. In addition, life consisted of simple microbial species, mostly bacteria, for the first 2 billion years. So if we found a very complex life form that was older than this, surely this finding would falsify the theory? Yet, there is one logical reply biologists who were reluctant to abandon the theory could make. They could claim that life on earth might have started earlier than we thought; perhaps it started 4.5 billion years ago. And then a catastrophic event, such as a meteorite hitting the earth, or a worldwide disease, wiped out everything (they already appeal to catastrophic events to explain gaps in the fossil record and radical changes in species, rather than modify or abandon the thesis of natural selection). They might argue that after the catastrophic event life started *again*, and the fossils we now have pertain to its second appearance; the anomalous evidence comes from a fossil from the *first* period of life. This might be possible logically, but otherwise would have no evidence to support it. Suppose we discovered lots of fossils that were very old but very complex: which view would this support, the view that evolution had at least two cycles, or the view that it is not true? It seems that it would falsify the theory. So it looks like this fifth case does in fact show that the theory of evolution can be falsified in exceptional circumstances. These circumstances may be too exceptional for some people, who might insist that from a practical point of view no matter what is discovered through our ordinary everyday observations of nature—no matter what we find—it can be pressed into the task of confirming, or at least as fitting into, the theory of evolution. This aspect of the theory has worried many. Although it is falsifiable in principle some might say, it is not really falsifiable in practice.

This brings us to the matter of theory change in science, an issue alluded to in Chapter 2. We need to remind ourselves that theory change in science comes slowly, and for good reason. Scientists, once convinced of the truth of a theory, are usually very reluctant to give it up because it changes the whole way they have to look at things (and even the way those of us who do not practice science look at things). This change can be unsettling. In addition, it is good practice for science to wait until the community of scientists has checked a theory over a period of time before abandoning one theory and embracing a new one. This is a very complicated topic, that would take us away from our concerns in this book, but we do need to note that the history of science is full of examples of theories that were once accepted as true (even as fact!), but that were later abandoned (some examples are: the humors of medieval medicine, the effluvia of early theories of static electricity, the existence of phlogiston as an element released in combustion, the electromagnetic ether that was supposed to surround the earth). This has chastened some, especially those who study the philosophy of science (but interestingly not most scientists), to be careful about claiming that our current theories are, in fact, the true ones.

We might use an analogy from the philosophy of religion about God's knowledge to help us understand this point. We can represent God's knowledge as a very large circle, and then place a much smaller circle inside this circle. The small circle represents our knowledge. It is very inferior to God's knowledge. Applying this model to our present topic, we can represent the complete body of possible (objective) scientific truth about reality by the very large circle, and the much smaller circle within the larger circle represents the present state of our scientific knowledge. This metaphor indicates that our knowledge of the realm of the physical is very small indeed; that there is much more to be discovered. This would leave plenty of room for us to find out that most of our current theories, like many theories of the past, while useful and practical now, are in fact false, perhaps even quite wide of the mark. This is in contrast to a second possible reading of the circles that depicts the circle inside the large circle as itself very large, almost touching the inside edges of the large circle, to convey the point that our current knowledge of the physical realm is vast and almost complete.[29] While there is still work to be done, according to this second model, there is not much left to discover; there will be no more major revolutions in science (modern cosmology explains most of the physical universe, evolution explains the development of life, genetics explains the chemical basis of life, etc.). It is difficult to know which of these readings of the circles is the correct one. Nevertheless, the metaphor should give us pause about claiming too much for our scientific theories, and also about becoming too attached to any particular theory as the only way that we can look at things, notwithstanding the argument that we must work with the best theories we have at any given time, and that at the present time, this would include the theory of evolution.

Although we can't pursue this question fully here, it is also interesting to consider what would happen if we came to the conclusion that the theory of evolution is false, if we no longer accepted the theses of common descent and natural selection? This would have a large effect obviously on paleontology. It would also mean that we would not be able to explain the origin of new species; we would still have this problem. But it would not affect so much our work with DNA and in the area of microevolution, or our work in fighting diseases, in terms of mutation, diet, climate, etc. Discoveries with regard to fossils, the nature of DNA, and mutations would all still be true, and so the work involving them would continue unabated. It would be at the more theoretical level that we would have problems with regard to the origin and development of life forms and species. Indeed, if we discovered that evolution was false, it might have more interesting implications in philosophy and theology than it would have in science! This brings us to the next chapter where we turn directly to the reaction to the theory, and to the implications of the theory.

4

Evolution: Reactions and Implications

It is simply a fact about the theory of evolution that it was controversial right from the beginning. We will see that there were different reasons for why this was so as we consider various reactions to the theory. The reactions of different groups and interested parties are often closely related to some of the perceived implications of the theory for matters outside of science. Many recognize that the theory raises substantive questions concerning chance and order in the universe, the origin and development of species, design in nature, the role of God in creation, the relationship between science and religion, not to mention the overall place of man in the great scheme of things. So it will be necessary also to consider these matters by means of a further examination of some interesting questions that arise concerning the implications of the theory of evolution. In this chapter, we will consider questions concerning the general reaction to the theory, whether there could be said to be different types of evolution, how the theory is extended to apply to other questions not within its domain, such as the origin of life, and whether evolution could be said to be an ideology. In the second part of the chapter, we will draw out various challenging and interesting implications of the theory in the areas of philosophy, religion, and morality, and the general area of worldviews. The rest of the book will then provide a more detailed discussion of these implications.

Overview of some reactions to evolution

It is appropriate to offer a few reflections about the general reaction to the theory of evolution from various quarters. Although, we are more interested in contemporary reactions than historical views, we will also canvass briefly in this section some of the early reactions the theory elicited. Responses to a theory are often quite revelatory about its general nature, and the way a theory is greeted and appraised by interested parties can also provide a good insight into its perceived broader implications. We can identify briefly in this section five different types of reaction to the theory. The first is what we might call the reaction from the point of view of science. This concerns not just the reaction of those working within biology and other scientific disciplines but the reaction from the point of view of simply regarding evolution as a scientific theory, appraising it as a scientific theory, leaving all other considerations aside. From this point of view, there was some initial skepticism, as there is when any new theory comes along that is attempting to displace other views. This is particularly the case when the theory is of a radical nature, like evolution. Scientists are usually reluctant to embrace a new theory right away, and often treat it with skepticism and hostility that borders on dogmatism (for example, in physics Michael Faraday was ridiculed by all the leading scientists of his day for first suggesting what we now know to be the case—that there is a close relationship between electricity and magnetism).[1]

The debate lines (perhaps we should say battle lines) were drawn immediately in the reaction to Darwin's theory. While it was warmly welcomed and championed by atheists, materialists, and liberal religious believers, it was greeted with disbelief and hostility by orthodox religious believers, and many others, including many scientists, who agreed with some of Darwin's conclusions, but rejected others. Distinguished contemporaries of Darwin, such as Richard Owen and Charles Lyell recognized that his theory was a major contribution to the conundrum of the origin of the species, but disagreed with various aspects of his argument, especially the notion of common descent. The American scientist, Asa Gray (1810–1888), welcomed the theory, and arranged for the *Origin* to be published in America, but he thought that evolution must be guided by the Divine Intelligence. Gray also rejected Darwin's claim, as did Lyell, that the higher capacities of man could come about by means of natural selection. The theory also proved controversial because it was regarded by many as an attack on religious belief, although Darwin claimed he had no such intentions. Yet many criticized him for wading here and there into questions about religion, about divine providence and the divine intelligence operating in nature, when he should be restricting himself to what the scientific research actually showed (just as we are often similarly critical today of biology textbooks for bringing larger issues into the discussion, thereby going beyond science). In short, the initial reaction very much resembles the current reaction, and it was often hard then as it is now to separate genuine scientific agreement or disagreement concerning the theory from the motivations of those scientists, religious believers, atheists and other thinkers who were reacting to the theory from the point of view of its place in a broader debate.

However, it is nevertheless true that with the passage of years, even in Darwin's own lifetime, and more so today of course, the theory of evolution came to be accepted, first among biologists, then within the broader discipline of science, and eventually even within the larger intellectual community. This means (ideally) that those who are closest to, and most familiar with, the evidence, such as evolutionary biologists, think that it is convincing. It is perhaps the case that other scientists further away from the subject (biologists who don't work in evolution, in addition to chemists, physicists, etc.) accepted the theory on the authority of those who worked on evolution and who were convinced, and then other intellectuals followed suit. This is often how it is with scientific theories, especially complex, abstract ones such as evolution, relativity theory, and the big bang theory; many of us are so removed from the evidence and details of such theories that we assent to them on the authority and trust of those who are experts. So there is nothing wrong with this practice in general (though if a theory becomes controversial, like evolution, it is perhaps no longer sufficient to accept it on authority). But this first type of reaction is that scientists and other intellectuals, who ideally are motivated by reason and evidence, gradually accepted many aspects of the theory as being supported by reasonable evidence (though this claim is not itself uncontroversial, as we will see at the end of this section).

The second type of reaction is that of those atheistic secularists who saw in evolution a possible way to attempt a justification of their atheism, and so they jumped on the theory as their savior. As Alvin Plantinga has noted, if you are an atheist (or naturalist) and you are called upon to justify your view that all of reality is physical in nature, consisting of some configuration of matter and energy, especially as this might pertain to the human person, evolution "is the only game in town." So you would be very quick to claim evolution as your own, and to appropriate it for your own purposes. A case in point was Darwin's friend Thomas Huxley (1825–1895) who immediately embraced the theory as a way of critiquing religion. In fact, Huxley was quite deliberate in adopting evolution as a way of supporting a position he called "scientific naturalism," the view that science provides the *only* reliable knowledge of nature and human beings. The aim of Huxley and his supporters was not so much to extend scientific knowledge, but to create positions and influence for themselves, or as historian Frank M. Turner has put it, "to expand the influence of scientific ideas for the purpose of secularizing society rather than for the goal of advancing science internally. Secularization was their goal; science, their weapon."[2] Science being used in the service of politics is not just a recent phenomenon.

This is one of the reasons the theory was controversial from the beginning. Everyone, supporters and critics alike, immediately saw the implications it might have for the larger debate between atheism and religion. Many responded to it accordingly, and failed to respect the distinction between regarding evolution as a scientific theory and co-opting it in order to press it into the service of atheism. This point is not only of historical interest, but is an ongoing issue, because, as secularism becomes an increasingly influential view in modern culture, there is now a pressing need for contemporary thinkers, such as Daniel Dennett and Richard Dawkins, to justify their position with arguments. These thinkers are looking around for a theory to support their naturalism, and evolution seems to fit this role.[3] And if it turned out that it was true, but did not particularly support any form of atheism, or if it turned out not to involve significant random elements, they would regard such consequences as major problems, and so they are reluctant to consider these options, and are often tempted to shut down discussion. As we have already noted, the careful thinker about evolution and religion must be aware of this general approach, and its pernicious overall influence in the debate. Not only has this approach done much to confuse the issues, but one can make a case that it does not even help the cause of science education. This is because it can make people hostile to science because they think, wrongly but perhaps understandably in certain cultural contexts, that science is a form of atheism, or at least that modern science is atheistic in practice and outlook (a claim advanced particularly by supporters of Intelligent Design theory, and one that is not far from the truth, at least in some circles of scientific practice). Unfortunately, this perception may discourage younger scholars from pursuing work in scientific fields.

A third reaction requires us to look at the matter from the religious side, especially from the point of view of creationism. Just as certain thinkers hostile to religion, like Huxley, recognized that they could exploit the theory of evolution to attack religion, so did the creationist movement in general interpret evolution as a threat to their particular religious beliefs. The history of this view has been interestingly documented in the work of Jon Roberts and Ronald Numbers, among others, and it would take us too far afield to explore this fascinating topic in detail.[4] But a range of thinkers, including theologians Charles Hodge (1797–1878), Albert Barnes (1798–1870), and James McCosh (1811–1894), and scientists who were also theists, including Gray, Louis Agassiz (1807–1873), and St. George Jackson Mivart (1827–1900), offered a variety of critical reflections on evolution, some harsher than others. The scientists in particular tended to appraise the theory naturally enough from a mostly scientific point of view and raised some of the familiar problems, such as the gaps in the fossil record, the lack of evidence for natural selection (along with the deeper problem, acknowledged by Darwin, that perhaps the process could not be empirically verified even *in principle*), the inability of the theory to account for the higher properties of man, and so forth. Several questioned whether evolution was even a properly formed scientific theory, and others criticized what they saw as special pleading, especially by Darwin, with regard to the evidence; for example, interpreting the presence of certain fossils as support for the theory, but regarding the absence of important fossils as not counting against it.

Many religious thinkers adopted a blend of scientific criticism and theological censure. The latter included the view that the theory contradicted the Book of Genesis account of creation, that evolution undermined the special place of man in nature, that evolution was motivated by atheism. Moreover, it would be impossible to provide foundational support for the objective moral order on such a theory. Many strands of religion, especially of Protestantism, felt threatened by the theory, not so much for reasons having to do with its general implications for religion, but because they saw it as a specific threat to their biblical literalism, especially to their reading of the creation story in the Book of Genesis, which they felt was directly contradicted by evolution. This view was well expressed by Barnes who held that "the truths recorded [in the Bible]... have their origin directly in the mind of God and have been imparted by him to the minds of the writers by a direct communication; ... they have been so guided by the Holy Spirit as to be preserved from error ... the Bible is ... one book, whose real author is the Spirit of God."[5] Rather than being guided by the judgment of the scientific community with regard to evolution, and trying to see how evolution might be compatible with their particular religious perspective, many theologians adopted a confrontational approach toward the theory, much like the naturalists adopted toward religion; as Hodge noted, "the Bible is to the theologian what nature is to the man of science; it is his storehouse of facts."[6]

This approach made the creationists reluctant to look at the evidence in a disinterested, dispassionate way, and many seemed to adopt a position against evolution almost in principle, one that could never be dislodged by any appraisal of the evidence. This reaction perhaps can be extended to cover the view that some may have been afraid of the implications of evolution for various aspects of religious belief (for example, for theories of biblical interpretation), and so rather than face these problems head on, they rejected the theory and avoided facing up to their intellectual responsibilities. There is some evidence in Roberts's and Numbers's research to show that this view motivated some creationists to reject outrightly evolution without giving it much of hearing (though the whole story is vastly more interesting and more complicated, because there were also many conscientious thinkers who remained unconvinced by evolution, but who engaged honestly with the questions it raised for religion).[7] We continue to see some of this reaction to the theory today, with some religious denominations rejecting it as a matter of policy, and being unwilling to consider it simply as a scientific theory, refusing to look at the claims of the theory and where it might lead one way or the other. This overall approach to the topic has given rise to a confrontation not only between creationism and secularism, but between creationists and other religious believers, and it is quite sad to see churches displaying signs with slogans like "We Teach Evolution" as a way of attacking other churches! Sometimes it seems that some religious denominations side more with the secularists than they do with their fellow religious believers!

This brings us to the fourth reaction, which we can treat somewhat more briefly. In addition to those who look at the theory as just another scientific theory, and those who welcome it primarily as a way of critiquing religion, and those who reject it because they interpret it as having negative implications for religion, there are those who accept the theory and who try to reconcile it with religious belief. This group accepts the theory because they think the evidence for it is fairly convincing. This group can itself be divided into two subgroups. The first subgroup consists of those who think that at least some aspects of the theory are warranted by the evidence, and then attempt to see how it can be reconciled with religious belief, and how we can iron out any anomalies that appear on the surface (for example, concerning whether human beings differ in degree or in kind from other species; or how chance and design can work together in the process). This subgroup regards evolution as raising tricky problems for religious belief that require a good bit of thought to resolve. It is not that these problems can't be resolved but they would require us (perhaps) to rethink some of our theological approaches and positions. In this category, I would place the views of thinkers like Keith Ward, Kenneth Miller, Paul Davies, and Richard Swinburne, among many others.[8]

The second subgroup was already referred to in Chapter 1. This subgroup positively welcomes the theory of evolution into their theology, seems to be glad that a theory like this is true, and seeks to then explain their religion in the light of it, to allow evolution, if you will, to define their whole approach to many theological and ethical matters. Why would someone welcome the theory so strongly? It can't just be because they find the evidence convincing, because the evidence alone does not determine one's reaction to a theory. The answer seems to be something that the new view makes possible. I think this group sees evolution as a way to rethink (i.e., to critique) much of traditional theology and philosophy on issues such as the nature of God, the nature of revelation, the place of man in the universe, and the nature of ethics. In short, there is often a theological (and perhaps moral) agenda at the back of a person warmly welcoming evolution, because it is an opportunity (an excuse even) to critique other theological views, which they have already rejected. Another way to express this point is that it is a way of finding further evidence, so they see it, in support of a view toward which they are already strongly inclined. I do not say that those who positively welcome evolution are all motivated in this way, of course, but it is an important part of the story that some religious believers *like* the theory of evolution because of their prior theological views, which are often at odds with traditional theological (and moral) views. This is particularly true I think of the movement known as process theology (referred to in Chapter 1). Thinkers in this tradition would include Arthur Peacocke and Ian Barbour.[9] (I will return to Peacocke's views in Chapter 7.)

I want to note one last type of reaction to the theory of evolution that is subtly different from the last two mentioned, but perhaps more interesting. It is a reaction that one finds among intellectuals who accept evolution. Some thinkers accept the theory in general outline, but nevertheless believe that it faces several problems. It is a very significant point about the theory that there are a large number of intellectuals across many disciplines who accept its broad outlines, and believe that there is some good evidence for these claims, and yet who think that the theory still faces some serious problems. These problems range over a number of issues: gaps in the evidence; "just so" or fictional stories (as noted in Chapter 3) to illustrate claims of natural selection in the absence of real evidence; aspects and features of life that evolution cannot explain; the cavalier approach many scientists have to the evidence, including their reluctance to describe evolution as a "theory," and their willingness to ridicule anyone who questions it (a point also emphasized by Plantinga). All of these points leave many uneasy about parts of the theory, even if they accept it as being broadly true (meaning that they might accept common descent, that some natural selection takes place, and that evolution can explain many things but not all that is claimed for it). I have met a very high number of very well educated people, including many scientists (most of whom have PhDs) who hold these views, and I regard this phenomenon in itself as very good evidence that evolution still has work to do. I do not regard people who hold these views as being biased, stupid, backward, afraid to consider the evidence, or not as enlightened as scientists! Indeed, they are among the most intelligent, reasonable people I have met, and given this, it suggests to me that there is still a number of questions regarding the claims of evolution and the evidence for these claims that troubles many thinkers of goodwill. And so it is a very important part of the story of the reaction to evolution to acknowledge this view, important to appreciate that many are troubled by the evidential gaps in evolution. This fact also cautions us not to paper over these gaps in our enthusiasm for a theory toward which we might be very inclined, perhaps for some of the reasons we have seen above, not all of them having to do with a rational appraisal of the evidence, or a desire to promote good science.

Is evolution an ideology?

The Reformed Protestant philosopher, Alvin Plantinga, is well known for his criticism of the theory of evolution, and especially for his claim that the theory serves as an ideology for many people. Plantinga argues that evolution is an "idol of the tribe," as he puts it, the *foundation* of a general worldview that many current academics share, especially philosophers, those in the hard sciences (particularly biology and physics), psychology, and some other areas of research.[10] Recall that although an ideology may be considered in a descriptive way, it also has a prescriptive meaning that today is usually understood in a negative way. The descriptive meaning is that an ideology describes one's worldview, one's philosophy of life, one's main beliefs about reality, ethics, and the meaning of life. In this sense, everybody could be said to have an ideology, or at least to live according to an ideology. But there is a prescriptive connotation as well, which today is often understood to mean that one is passionately committed to one's philosophy of life, does not entertain real criticism of it, holds it dogmatically, is quite intolerant of those with opposing or differing views, and is often trying to convert others to one's worldview, sometimes using unfair methods that may involve bias, dishonesty, prejudice, logical inconsistency, subtle or devious tactics, and so forth. When Plantinga describes evolution as an ideology, he intends it not just in the descriptive sense, but also in the prescriptive (or pejorative) sense.

Plantinga believes that many thinkers today, especially in our universities, and perhaps especially in the disciplines of science and philosophy, reject the religious view of the world, and replace it with an atheistic view, which they defend by appeal to modern scientific theories, especially the theory of evolution. We recall from Chapter 1 that this alternative view is called secularism (or naturalism), and that it is often confused with science. Plantinga's point is that for these thinkers, since secularism is their alternative worldview to religious belief, and since they often back up their worldview mainly by appeal to evolution, then evolution becomes *a very important theory* indeed for them. The thinkers he has in mind include Dawkins, William Provine, and Dennett, who vies with Dawkins in his disdain for religion and in his irrational exuberance for the explanatory power of evolution. Dennett claims dogmatically that "An impersonal, unreflective, robotic, mindless little scrap of molecular machinery is the ultimate basis of all the agency, and hence meaning, and hence consciousness, in the universe."[11] Plantinga rightly regards this as an irrational claim, but it does illustrate that for these thinkers evolution becomes one of their main ways of *thinking about reality*, and of justifying their most important beliefs. It becomes, as Plantinga puts it, "an idol of the contemporary tribe; it serves as a shibboleth, a litmus test distinguishing the ignorant and bigoted fundamentalist goats from the properly acculturated and scientifically receptive sheep ... among the secularists evolution functions as a myth ... a deep interpretation of ourselves to ourselves, a way of telling us why we are here, where we came from, and where we are going."[12] Evolution is no longer simply a scientific theory that explains a few natural processes and offers a few new facts about which we might have interesting discussions regarding the evidential support, and future research. It is now a theory around which many try to justify their whole worldview, and so Plantinga claims these thinkers tend to hold it as an ideology. This means that they are prone to overstating the evidence for evolution, prone to claiming more for the theory than it claims for itself, prone to co-opt it and press it into the service of atheistic naturalism. They are therefore also strongly tempted to ridicule anyone who questions the theory, not because it cannot be questioned, but because they have so much invested in it that they cannot allow an honest discussion of the larger questions (such as whether evolution shows that there is no design). According to Plantinga, they are likely to suppress or discourage an open discussion of the theory, likely even to exaggerate the evidence for the theory, or exaggerate what the theory can explain, *because there is so much at stake*. It is in this sense, Plantinga argues, that evolution is an ideology.

I think that Plantinga is right about much of this, and anyone who has read a good deal of the literature on religion and evolution over the past twenty years would find it hard to disagree with him that many use the theory in this way. It is obvious that for many it functions as much more than a scientific theory; that it really serves as an ideology to which they seem to be passionately committed, which they appear to accept on faith, as it were, and hold very dogmatically, and about which they are reluctant to raise foundational questions. Regrettably they are also quick to deride anyone who dares to ask a critical question, a tactic that makes many reluctant then to raise criticism in public. There is no better recent illustration of this phenomenon than the somewhat hysterical reaction to two recent books that raise critical questions about some aspects of evolution, both written by highly respected, prominent (atheist) American philosophers, Thomas Nagel and Jerry Fodor, and by Italian scientist, Massimo Piattelli-Palmarini.[13] It goes without saying that not everyone who believes in evolution holds the theory as an ideology; in fact, probably a minority regard it this way, though Plantinga thinks it is more like a majority in the university setting (especially among those who don't work directly with the theory in science), and that the general cultural influence of this group is growing. Of course many (and we hope the majority) simply accept evolution as a scientific theory, consider it the way we would consider any other scientific theory, and don't worry too much about its implications. Many of this latter group would also hold that it is compatible with a number of different worldviews, both religious and secularist, or that it has no special implications that we need to worry about for any particular worldview (we will return to this matter below). So for this group it does not function as an ideology.

However, it is important to be aware of Plantinga's point in the general cultural discussion in which the theory of evolution plays a formative role. We do need to be aware that evolution does function as an ideology for some, and we should be alert when this is the case, because it will influence our appraisal of their arguments. It is critical in other words to recognize and to be on the lookout for the distinction between discussing evolution purely as a scientific theory, and the evidence for it, in just the way one would discuss any other scientific theory (e.g., photosynthesis), and evolution serving as an ideology in the prescriptive or pejorative sense. One should be rightly suspicious of this latter approach because it is unlikely that it will involve a *dispassionate analysis* of the theory, and will likely feature an exaggeration of the implications of the theory in the service of a secularist ideology. We need to be especially aware of sophistical approaches like that of Dawkins and Dennett, who appear to have trained themselves to regard the central philosophical questions arising out of these topics as silly, but this subjective response does nothing to reduce the importance of these questions, or help us with the answers.[14] Their approach is to pour scorn on, and to express and invite incredulity toward, not just religious answers, but even the fascinating questions raised by religion, as a substitute for argument. So our answer to the question "is evolution an ideology?" is that it is not an ideology in itself; it is simply a scientific theory. However, it is often *adopted* as an ideology by some, and it is very important to know when it is being used this way. Unfortunately, its appeal as an ideology is becoming more widespread, and we even see it creeping into science textbooks, especially biology textbooks.[15] It is also sometimes the case that various scientists are not careful to keep the theory of evolution distinct from an ideology that it might be used to support (including perhaps their own ideology), and often this confusion creeps into their public pronouncements about evolution, and about science more generally.

Are there different types of evolution?

It may seem that since there is only one theory of evolution, there can be only one type of evolution, and so the answer to our question must be no. However, the concept of evolution is used in several different ways, and is applied in several different areas, not all of them having to do with biology, and for this reason it can often lead to confusion. In many discussions involving evolution, for example, it can be helpful to distinguish early on what type of evolution is intended so as to minimize confusion with the official biological theory of evolution. So it will be helpful to take a few moments in this section to distinguish between various types of evolution that one might consider, all of them inspired by the biological theory of evolution, but each of them emphasizing some particular topic or feature that may (or may not) be part of the official theory.

The first and main type of evolution, and what perhaps should always be intended whenever we use the term (but is not always intended, of course) is the official theory proposed by Darwin and developed by modern biologists working simply as biologists (and not as atheists, secularists, religious thinkers, or someone else with an agenda that they are using science to support). These biologists should be working also just with scientific evidence, and with what they think such evidence shows. This is the theory outlined in Chapter 2, the one that appeals to the central concepts of evolution, such as common descent, natural selection, survival of the fittest, micro- and macro-evolution, to explain the change and development in various species in nature over time. These concepts are supported by appeal to scientific evidence, as we have seen (e.g., the fossil record and DNA evidence), and this is supposed to be ordinary common sense evidence that we can know through reason and sense experience. Perhaps our optimum scenario would be if this were the only meaning of evolution in our culture, and meanings that were not bound up with outside worldview issues never arose! This optimum scenario would have led to less misunderstanding, less confrontation, and less problems all round!

However, we know that other meanings have emerged over time in the general discussion, and evolution often means different things to different people. A second type of evolution might be one that co-opts evolution as a way to attempt to explain the *origin* of life, and not just the development of life. The theory of evolution proper is a theory about the development of life, not its origins. It purports to explain how life developed *after* it originated. However, the development of life inevitably gives rise to the question: how did life begin in the first place? And some argue that evolution might be a way to tackle this question as well. The research project that addresses this question is often called abiogenesis, the hypothesis that (organic) life originated out of non-organic chemical compounds on earth about 3.5 billion years ago. The speculation is that perhaps at the right time in cosmic history many of the ingredients thought to be necessary for life, such as ammonia, hydrogen, and methane, were all present in the same place at the same time, and something gave them a spark that triggered off a type of "life force" through them that gave rise to a living organism. It is often speculated that this "something" may have been an electrical charge, perhaps from a lightning strike. Some proponents of this theory then suggest that this first life form split into two, continued to divide, eventually leading to simple species, that gradually became more complex, and this process continued for billions of years (as we saw in Chapter 2), eventually leading to the emergence of *Homo sapiens*. These claims give one an idea of not only how radical this view is—that life began this way, and also that it evolved to such complex heights—but also how far-fetched such claims are without being backed up by some very good evidence. (This is why many people, as we noted in Chapter 3, look at claims like these and the {lack of} evidence for them, and are not convinced.) But our point is that this theory about the origin of life is not an official part of the theory of evolution; it is an add-on to the theory by those keen to understand not just the development of life, but also its origins.

Questions about the origin of life are legitimate questions, of course; abiogenesis is an attempt to develop a scientific answer to these questions. Even though not part of the official theory of evolution, this thesis is now presented in many expositions of evolution and in many biology textbooks as if it is part of the official theory. More importantly, it is often presented as not just a speculative hypothesis, but as a serious theory with the implication that there is some good evidence for it, even though this is not the case.[16] In fact, abiogenesis has made little progress since the Urey–Miller experiments of the 1950s at the University of Chicago. These were experiments where scientists Harold Urey and Stanley Miller worked with nonliving, inorganic components (those thought to be present on earth when life began) and tried to generate something organic from them. They were successful in producing amino acids, which are an important part of the structure of life, sometimes called the building blocks of proteins, but not life itself. However, since then little further progress has been made, and some think the whole process is fatally flawed.[17] Yet biologists often present this experiment as evidence that life originated this way, and this move simply leads to further confusion surrounding the theory of evolution.

The third area to which the concept of evolution is often applied is the universe itself, however silly this may sound. That is to say, in addition to the question of how life began, we have to ask how did matter begin? Even if we granted that life began in the naturalistic way claimed by some supporters of a purely evolutionary account of life, it would still be necessary to explain where the ingredients came from that gave rise to life. Sometimes naturalists and some evolutionary biologists, such as Richard Dawkins and Carl Sagan, attempt to apply the notion of evolution to the universe itself, an approach called "cosmic evolution."[18] Sagan argues that we can describe the universe as evolving from the time of the big bang onwards in the sense that each present state of the universe is causally produced by the preceding states. This process produces all of the galaxies, including our own Milky Way, eventually leading to the formation of earth, with its specific characteristics. Sagan then appeals to biological evolution to propose a speculative theory of the purely naturalistic origins of life and its development on earth. The notion of cosmic evolution is an interesting idea, and it might help us in some way to think about successive states of the universe. (It will also lead to Sagan's claim, which we will consider in detail in Chapter 7, that human beings are made of "star-stuff.") One can see in such a notion, however, the way in which some naturalistic thinkers lose the run of themselves when talking in this way about the general concept of evolution! They begin to *imagine* that the theory can explain everything, including the structure of the universe, the origin of life, the laws of science, even the origin of matter![19] This theory of "cosmic evolution" is another example of naturalists attempting to use current scientific theories in order to address those ultimate questions that we would normally turn to religion to explain, yet the naturalistic theories rarely address the ultimate questions, as we will see in Chapters 7 and 8.

A fourth application of the concept of evolution concerns the unique features of man. It is one thing to wonder if evolution, specifically the mechanism of natural selection, can explain the origin and development of the physical body along with its characteristics, say in a specific species like whales, or alligators, or human beings, but it is quite another to suggest that it can explain higher level faculties that we find in man, such as those of consciousness, reason, free will, and moral agency. Officially the theory of evolution is concerned with explaining the physical characteristics of species, e.g., their skeletal structure, digestive system, and the origin and development of their specific organs, and (let us not forget) *every other single feature they possess that is of a physical nature*. In principle, the theory of evolution is supposed to be able to explain all of an organism's features. It is obvious how notoriously difficult this task is, hence the presence of so many "just so" stories in the literature on the theory. But it is another thing when we come to features, especially of human beings, that many philosophers, scientists, and other thinkers have argued do not belong to the physical realm, such as consciousness, reason, free will, and the human soul. Can evolution explain these phenomena?

The official theory did not originally claim these phenomena to be among those features that it purported to explain, and early theorists, including Darwin, were often reluctant to discuss them, except perhaps obliquely, because it was obvious that the topic would lead to further controversy. In recent times, the theory often works with an assumption that evolution may be able to explain them, or perhaps at least that they should be part of its research agenda (just like the origin of life should be). This has gradually morphed into the *assumption* that evolution *can* explain them, even though these phenomena present a very difficult problem. But again we must point out that this is not part of the theory proper, and that it requires a bit of a leap to claim that evolution will be able to explain complex features of *Homo sapiens*. It is important to note that if a thinker begins with the view that the conscious mind is really a physical thing that is produced by the brain, this is an assumption about the relationship of body to mind that can drive a research agenda in evolution. But this is not the same as saying that the evidence for evolution shows that the mind originates from the brain, for the simple reason that it does not show this! This has been a notoriously difficult area for philosophers and scientists for centuries; one can, to be sure, make a commitment to a naturalistic view, but it is just that, a commitment, and does nothing to settle the question one way or the other. Brian and Deborah Charlesworth have written that "... there is little doubt that all forms of mental activity are explicable in terms of the activities of nerve cells in the brain"; adding (and seemingly uninterested in, or unaware of, the controversial nature of these claims), that "most people would not regard a newborn baby as conscious," and that even though we know little of the evolution of human mental and language abilities, "there is nothing particularly mysterious in explaining them in evolutionary terms."[20] The Charlesworths's confidence comes from a prior commitment to naturalism rather than from actual scientific evidence. If one does not share their commitment to naturalism, then there is much more room for doubt. The Charlesworths and others must also be careful to recognize the danger that they are coming very close to begging the question at issue because one cannot assume that naturalism is true if the question of the nature of consciousness is yet to be decided; and if one *commits* to naturalism before the nature of consciousness has been resolved, one must recognize that the commitment is provisional. (We will raise further problems for the thesis of naturalism in Chapter 7.)

Our main point is that the theory of evolution proper does not provide any explanation for these phenomena, *nor is it likely to*. If one operates with the assumption that evolution must contain an explanation of them, one must recognize that this is an assumption, and not a scientific fact. But sometimes the theory of evolution is presented as if these phenomena are part of its purview, thereby creating the incredibly misleading impression that evolution can explain all of the difficult phenomena of human life, including free will and moral values. When in fact all we have is a lot of almost useless and certainly false speculation about the origin of these phenomena, which has no basis in scientific evidence. So it is very important to distinguish this understanding of evolution from the theory proper. We should also note that while we are focusing in this book on the difficulties evolution might pose for religion, and how we might address them, we have just been talking about the considerable difficulties that evolution also poses for secularism (naturalism), and that the prospects for solving them do not look very bright. This focus on complex human phenomena concerning evolution, and religion and naturalism, brings us directly to the crucial topic of the implications of evolution for philosophy, religion, and morality.

The implications of evolution

It is now time to draw out more fully and in more explicit detail the implications that evolution might be said to have in the area of religious belief. Many people on both sides of the argument—the religious side and the secularist side—seem to agree that evolution at least appears to have a number of quite significant implications for the subject of religious belief, and for the subject of worldviews, ethics, and the meaning of life more generally. Indeed, these perceptions of the implications motivated many of the reactions we have just considered. Of course, whether it actually has such implications is sometimes itself the subject of dispute, so we must keep this in mind. Our task is to attempt to identify the implications that the theory raises, and then in the rest of the book we will elaborate some of the more significant implications, and consider and develop our answers to them. We can classify the implications of evolution into two general areas: theological and philosophical (including ethical), allowing perhaps for occasional overlap between them. But keeping these areas distinct is a good way of organizing the myriad questions and problems, claims and counterclaims that the theory of evolution gives rise to regarding different areas of life, history, and experience. Bringing order and discipline to the debate that surrounds evolution and religion is a difficult task in itself, and so this is also one of our main aims in this book, as we attempt to help readers navigate through the different pieces of this absorbing, but often quite challenging, topic.

Theological implications

Let us turn first to the various theological implications that evolution is often thought to have for religious belief. As noted in Chapter 1, we have in mind here Christian belief, since this is the form of religion that most Western readers will be mainly familiar with, and are more likely to practice. It is also Christianity that is usually the focus of discussion when we are talking about religion and evolution. However, we should note that many of the implications we raise concerning Christianity will usually apply in most religious traditions, as long as they allow a significant role for reason and science.

The first and most obvious question evolution raises with regard to Christian belief concerns Christian revelation. Christianity holds that God has revealed his word and his teachings on key matters (concerning salvation, redemption, morality, and so forth) in the Bible. Although this is not the only way he may have revealed his word, and many thinkers believe that we can also work out aspects of Christian teaching by means of the light of natural reason and thinking about the natural world (often called the realm of general revelation), nevertheless most Christian thinkers, whatever their denomination, accord a very large element of significance to biblical revelation. Evolution is thought to undermine revelation because it appears to support by means of scientific evidence various facts about nature that seem on their face to be inconsistent with the biblical accounts, and so the question arises as to whether evolutionary theory actually undermines biblical claims. The most obvious case is the creation story in the Book of Genesis, which seems to suggest that what some have called the "fixity of the species" view (the view most students of nature held before Darwin) is the true account of the origin of species. More specifically, the Bible indicates that God created the species fixed and intact, as it were, and so this would mean that the evolutionary account is not true. On the other hand, if the evolutionary account is accurate, then the biblical account could not be true. And so since the evidence appears to provide good support for evolution, this would undermine the biblical description of the origin of the species. Some have gone further and suggested that this might undermine the Bible in a more general way, in the sense that if the creation story is not correct, then perhaps other accounts of historical events, significant happenings and the experiences of important figures are not reliable either, in both the Old and New Testaments. One might reply to this line of reasoning by noting that perhaps the biblical writers used stories here and there to make deeper points, that they employed some literary license to make foundational points more evocative, clear, readable, immediate, etc., but at what point then does one begin to think that these are just stories, and not revelations from God at all? So evolution is thought at least to raise these questions, irrespective of where one might come down on them eventually.

There are however other implications for the Bible, in addition to these familiar claims about Genesis in particular, and about revelation in general. It is sometimes claimed that evolution shows that certain facts appear to be true about the species that would contradict the biblical worldview, at least as it had been traditionally understood. For example, evolution seems to suggest that human beings do not differ in kind but only in degree from other species (we will come back to this point below), and Christianity has always understood that human beings are in a different, more special category than other species, and indeed that human beings have dominion over other species. If evolution is true, it would seem to challenge what is regarded as a central claim of Christianity.

Looking at the topic from the point of view of how, if we accept evolution, we would reconcile it with religious belief, one question we would have to ask is why would God use the process of evolution as a principal means of creation? Is there perhaps some theological rationale behind evolution? Is it a morally better way of bringing about living creatures than by creating species intact, all at once, as it were, and if so, in what way is it better? (Some claim it is a better way.) Another important question we need to look at from the point of view of human reason and science is how specifically God creates through evolution? There are two main possibilities to consider. Did God set the process in motion and let it then develop by itself to produce the various species (and we would have to think about how such a process would work)? Or perhaps God created in the way some creationists and Intelligent Design theorists argue, by setting the process in motion and then occasionally intervening in creation at special points to bring about his desired ends (these interventions are sometimes called points of "special creation")?

A further issue that is related to the general question of design in nature concerns the fact that there is a lot of evil in nature and in human experience. The implication we want to point out here is why would God include evil in the process of creation, or allow it to enter creation? It might seem that this question, though a very important and challenging one for many believers, does not have much to do with evolution. However, some hold that the theory of evolution makes the problem of evil worse in various ways, and so it raises more urgently than before evolution the question of why nature would be filled with vicious struggles for existence between various species, pain and suffering, extinctions, disease, chance events, and so forth (we will return to this topic in Chapter 8). This brings us more directly to the implications evolution appears to have for our own species, *Homo sapiens*, especially for the questions the theory raises about religious understandings of human nature, the origin and purpose of human life, and morality, among other topics.

Philosophical implications

Philosophers have raised the question of whether human beings differ in kind or in degree from other species, but in a different way—by emphasizing important concepts and categories in philosophy. Philosophers have long pondered questions relating to the nature of human beings. Do human beings have a nature or an essence; an essential set of traits and characteristics? If so, are these merely biological or do they extend beyond biological explanations? These topics are related to the question of differences in kind and in degree. If we differ only in degree from other species, does this mean we have the same moral status as other species? If in kind, in virtue of which particular qualities (e.g., soul, consciousness, reason, free will)? These philosophical questions have obvious, very important religious implications. Although the theory of evolution does not deal with such questions directly, various thinkers argue that it nevertheless seems to have implications for many of them, implications that we must take seriously.

Some argue that the theory of evolution appears to contradict the view that human beings differ in kind from other species. If the evidence from common descent is correct, then it means that human beings are genetically related to not only their immediate ancestors, such as chimps, gorillas, and bonobos, but also to more distant species such as fish and birds, and bacteria and plants, all the way back to the first living thing. This is what the thesis of common descent claims, and so it is one of the key claims—perhaps the key claim—of evolution (along with natural selection, which is the mechanism by which one species descends from another, by which new species emerge). Although it does not follow of necessity from the thesis of common descent that human beings do not differ in kind from other species, some argue that the process of evolution seems to *suggest* this conclusion. It would be reasonable to conclude that if the same process produces all species, including us, then we are probably all of the same kind, all in the same category, only differing in our degrees of development, as it were. This argument might apply not just to obvious characteristics such as skeletal structure, length of stride, stomach capacity, and brain size, but also to features like mental faculties, language use, and intelligence. One might even suggest that evolution shows that species exist along a continuum in terms of intelligence: some have none (plants, bacteria, insects), others a little (dogs), some a bit more (chimps), and others have a great deal (us!). To use a computer analogy, *Homo sapiens* might have a Pentium III processor, a chimp a Pentium I, and so forth, but all species are essentially in the same class, rather than all other species being placed in one category, and *Homo sapiens* placed in a separate category by virtue of our special properties. So the theory of evolution raises these questions about human beings. It would naturally cross someone's mind who accepts evolution that human beings might be just the most advanced life form (no one denies that), but perhaps should not be placed in a different category, since all species are related to each other genetically, and came about by means of the same process. Another related question is whether it matters in any way that we differ in kind or only in degree? Proponents of traditional religious and moral views always insisted that we differ in kind, but if this turned out not to be true, what would be the consequences? So both of these interesting questions will have to be examined in the light of evolution.

Evolution raises other concerns about the coming into existence of various species, the nature of various species, and in particular whether these natures are fixed or not. In relation to these matters, it poses a few interesting but difficult conundrums. For instance, we can ask if the human species necessarily had to exist? The concept of necessity is an important one in philosophy, and it is often asked about the universe itself (did it have to come into being?), about the laws of the universe (could they have been different?), about mathematics (could mathematical relationships that apply throughout the universe have been different?), and in evolution about the various species that exist. When we ask if the human species is necessary, we are asking whether evolutionary history might have unfolded in a different way, whether different species could have evolved than the ones we have now. If evolutionary history had gone differently, might human beings not have come into existence? Could evolutionary history have gone differently, or is this highly unlikely, or perhaps impossible? Again the theory of evolution does not address these questions directly. Indeed how could it? The job of evolutionary theorists is to show how they think a particular species emerged, in terms of evidence, timelines, and so forth, and how it might be related to other species. It would be very difficult, perhaps impossible, for this kind of approach and evidence-gathering to show whether a species *had* to emerge or not, had to come into being. Nevertheless, it is a question suggested by the theory, given the way it says species did emerge, and so we have to consider it. And of course the implications are large: what if it turned out that our existence was a matter of chance; that we did not have to exist?

What would this mean for our view of God's intentions, God's plan, and our view of our special place in nature? So this is a very large question with many implications.

There is an extension to the question. In addition to asking whether our species as a whole is necessary or not, asking whether or not we had to emerge from the evolutionary process, we can also ask whether we had to have the particular nature that we have. Many philosophers in history (including Aristotle and St. Thomas Aquinas) considered whether human beings have a nature or essence that is objectively the same for all humans, a set of essential characteristics that all human beings share. This topic has implications for design, nature versus nurture questions, and morality. The theory of evolution may challenge the view that there is a human nature, or at least the view that any human nature is necessary, has to be the way that it is. I like to illustrate this question with the example of alien species from science fiction stories. In many science fiction movies and shows over the years, there is often a disappointment when one finally encounters the alien species (especially if the suspense in building up to their appearance has been intense!). The reason for the disappointment is that the alien species are often much like ourselves! They usually differ only in their outward appearance, that is, in their biological structure. But they are usually very similar, perhaps identical, in the areas of reason, mathematics, and language-structure, and they usually have similar moral beliefs! We might accuse the storytellers of lacking imagination but it is very difficult indeed to give alien species a different reasoning or logical system, a different mathematics or science. It is much easier to give them a different type of body, and a different language (but not language-structure), and perhaps a different set of moral beliefs (but not a different overall moral order) than us. But evolution raises this question: could we have evolved with a different body structure? The answer seems to be yes. Could we have evolved so that we had four arms, for example, or four stomach compartments (like cows, and great at a buffet!), or be one-eyed, or markedly hirsute?! The theory of evolution seems to say yes; if evolutionary history had gone a bit differently, many of our features could have been different. They are contingent, not necessary.

Suppose we extend the question and ask if we could have had a different type of intelligence? Could we have evolved with twice the level of intelligence on average than we have, or perhaps with a different order of intelligence altogether, including a different logical system (hard though it is to imagine what that would be like)? Could we have a totally different set of morals, and/or a different foundation for morality? These are hard questions, but evolution raises them. If the answer to them is yes, it would mean that those characteristics we now regard as part of the human essence are only chance occurrences; they are not necessary, just the type of nature we happened to develop. We can ask these questions too of other species. We can say that to be an elephant you must have a trunk, to be a giraffe you must have a long neck, to be a camel you must have a hump, but if evolution could have gone differently, then we might have elephants, giraffes, and camels without their usual characteristics. So indirectly the theory of evolution seems to be a threat to the idea of necessity in nature, to the idea of an essential set of characteristics that species must possess. (This would also have obvious implications for the question of whether it would be moral to interfere with animal or human genomes in genetic engineering, or to create chimeras, or to develop some type of artificial or synthetic life. Would modifying natures be such a problem, if it could be done technically, given the way evolution shows the "natural biological order" came about, and given that "the natural" could have been different, right down perhaps to the structure of trees and plants, and the chemical composition of water, let alone animals and man?)

A related question concerns the *future* of evolution. We usually look at the tree of life from its beginning up to today. But we can also try to imagine what the future holds in terms of evolutionary development, and not just in ten years, or 1,000 years, but in millions of years, or in 3 billion years. What will the tree look like? Of course it is impossible to say, but we are interested in two particular points: will *Homo sapiens* continue to survive (and perhaps evolve further?), or will we become extinct, with some other, more advanced types of species emerging in the future, taking our place, as it were? And what would this mean for our religious truth? Might it be the case that God's plan is for *them* (the more advanced species) to emerge as the main species, and not (as we have mistakenly thought) us?! We have good reason to think that we are supposed to be the main species because we are at the top of the evolutionary tree and because we have special qualities of consciousness, self-awareness, reason, free will, and moral agency. We also *understand* the process of evolution, and we have some measure of *control* over it (two *remarkable* achievements, that we will come back to in later chapters). (So we are not just at the top in the way birds might have been at the top at one time in the past—it is for this reason among others that we think we differ in kind from other species.) But might some other more advanced species come along in the future, and what would it all mean for God's plan, for the existence and place of our species in nature? These are all absorbing but challenging scenarios suggested by the theory of evolution.

Another important area in recent discussions concerning evolutionary biology involves the origin of morality. What are the implications of evolution for moral values, and the existence of the moral order (in addition to human reason and free will)? This is one area where the questions are just as difficult (indeed more so) for the secularist as they are for the religious believer. This is because if one holds to an atheistic naturalistic view, one must explain how our objective moral order could have originated from a purely naturalistic evolutionary process, one operating without any direction, a very difficult problem indeed. It is not clear how one can say much here beyond simply that morality evolved, just like language evolved. And this move would appear to undermine seriously the objective nature of the moral order because how could we claim that moral values are objectively true in some trans-historical sense, since if evolution had gone differently we would surely have a different set of moral values (just as different languages could have arisen, with different grammars, vocabularies, and spellings). Would evolutionary accounts of morality inevitably lead to a position of moral relativism in several senses: one sense would be that we might have to concede that, just as we could have evolved with a different biology, we could also have evolved with a different set of moral values; another sense is that what we now regard as immoral behavior could be justified as being part of nature, as being somehow useful from an evolutionary viewpoint, otherwise why would it have evolved? In addition, we can ask if evolution undermines important areas of Christian morality that seem to be in direct conflict with the competitive, cut-throat, selfish, mindless nature of the process. These areas include altruistic behavior, love of neighbor, sacrifice, Christian love, and the traditional view of the human person, which puts the emphasis on the social nature of persons and on the community, rather than on one's own individual interests. We will return to the topic of evolution and morality in more detail in a later chapter. But for now we can say that these are all challenging questions, and not just for the religious believer, but also for the secularist who will need to respond to them and who will have to explain the emergence of an objective moral order from a process that is itself (according to them) non-moral, impersonal, directionless, and governed by chance.

The final area of interest raised by evolution concerns the question of chance in the universe, a major theme of this book. The role of chance in the process of evolution is clearly raised by several of the points noted above. Indeed, the concept of chance is often lurking in the background in any discussion of evolution but is seldom defined carefully or subjected to detailed analysis. But many evolutionary biologists are clear in saying that the process of evolution involves quite a bit of chance. This kind of talk is very common in the literature on evolution, with evolutionary biologists frequently saying that evolutionary change occurs by chance, that things might not have happened the way they did and so evolutionary history could have gone differently, that there is a lot of randomness in the process, especially in the area of mutation, and so forth. Indeed, this view is now part of the official way evolution is explained, and taught in most biology courses and textbooks. Talk of chance thereby raises many of the scenarios we mentioned above: that species are not necessary but accidental, that they need not have come about at all, that they might have been different in their natures, that there appears to be no necessity driving the process of evolution. The upshot of all of this talk about chance and randomness is that it is often presented as a more general argument *against the existence of design in the whole of nature, and in the universe*. The claim is that if species came into existence in some random way, they cannot have been intended by a divine Mind; if their existence is due in large part to a series of chance events and accidental occurrences, they could not have been planned; if they could have emerged differently, they have no intended natures; if they have no intended natures, this would have very significant implications for our whole understanding of human life, including for the nature of morality. These are not necessarily questions that evolutionary biologists must answer; but they are questions that evolution, if true, identifies. This topic is also related to the question we raised above about the evil we often find in nature, and also, very significantly, to the question of whether it would be possible for God to *direct* the outcome of a process that still contains elements of chance, or would this be a contradiction in terms?

We have identified a host of very difficult questions that evolution raises in the areas of science, theology, philosophy, and morality. Many of them we cannot answer definitively, or in any kind of detail, and it may very well be that some of them have no *scientific* answer. The proof of the pudding on this matter would be in coming up with scientific explanations, backed up by good evidence, rather than offering speculative hypotheses. Nevertheless, we must consider in the chapters that follow how religious believers can respond to these implications, and how we can develop the argument that, given the truth of evolutionary theory, religion and evolution are quite compatible. This will require us in later chapters to consider carefully the various possible theological implications of evolution, and how we might respond to them. But the next two chapters will turn first to one of the key philosophical implications raised and one of the important themes of this book: the question of chance in the universe in general, and in evolution in particular.

5

Evolution, Chance, and Determinism

The roles of chance and randomness, and their opposites, necessity and determinism, throughout nature and the universe in general, as well as in the process of evolution in biology in particular, while raising a fascinating cluster of questions, are quite intricate, and also the subject of much uncertainty. Our aim in this chapter is to introduce these concepts in some detail, to explain them carefully, and to bring some order and clarity to our understanding of them. This will prepare the groundwork for our discussion in the next chapter for the ways in which chance and necessity might be related to the question of evolutionary development, in particular. I will argue in this chapter that there is much confusion surrounding these notions across a range of disciplines, but especially in biology, and that we *regularly* see chance and randomness confused with each other, randomness confused with unpredictability, and a carelessness in definition to such an extent that the issues at stake are not clearly brought out or understood. These confusions are perpetuated throughout evolutionary biology, and in biology textbooks. I will also argue in this and the next chapter that, despite opinions to the contrary, *there is no chance operating in our universe*, or to put this point in another way: that our physical universe (excluding free human actions, and any action by God in the world) is completely deterministic. Before I explain as clearly as I can what we mean by all of these terms and claims, and the important role they play in our specific discussions concerning questions of design and the occurrence of chance events in the universe, and of random events in biology, let us begin with a brief word about necessity and contingency.

Necessity and contingency

The question of whether any events happen *by chance* raises the question of the role of necessity and contingency in the universe. The concept of chance also raises questions with regard to the existence and nature of life, including human life. Matters of necessity and contingency have always been fascinating to philosophers because they involve asking a very interesting philosophical question: *which events and happenings in our universe, if any, have to be the way they are, and could not be otherwise*? Many philosophers have given attention to this question when thinking about various topics, including logic and the nature of conceptual thinking, mathematics, the nature of causation, the question of progress in nature, and the architecture or structure of the universe. It comes up again when we consider recent scientific theories, especially evolution. Are there any features of the universe that must be the way they are? Or could every feature of the universe have turned out differently if events had unfolded differently? *Could* events have unfolded differently? This latter question is obviously central and my answer to it in this chapter will be a qualified "no." But we might ask if, for instance, the laws of physics could have been different? Could the universe have developed in such a way that Newton's law that "for every action there is an equal and opposite reaction" does not hold true, and where there is a different mathematical relationship between material bodies? Could human beings have ended up in a universe in which a different logical system than the one we actually have obtains? For example, could we have ended up with a logical system in which the basic principles of logic are quite different from those we actually have now, where the following deductive syllogism is not true: all men are mortal, Socrates is a man, therefore, Socrates is mortal? Or a universe in which the force of gravity is only half of what it is in our universe? Might we have had a different mathematical system, one in which the area of a circle is not πr^2, but πr^3? Might we discover perhaps in the future that there is an alien species in some far-off galaxy that has a different system of science than ours; not just more or less scientific knowledge, or different technologies, but different basic scientific laws, reflecting a different set of relations between the elements that make up the universe?

Turning to ask similar questions about evolution, more specifically, we can ask whether a whole series of different species might have evolved than the ones that actually did evolve? Might human beings not have evolved at all, or might we have had a different lineage (say with elephant-like species rather than the great ape as our recent ancestors), or a different biological structure (six fingers on each hand perhaps)? It is important to be clear about what we are asking here. We are not asking if God might have designed the universe differently. God probably could have designed the universe in many different ways. We are asking: given our study of natural laws, logic, mathematics, and the natural processes that produce effects in the universe, *might all of these causal processes have produced something different from what they did in fact produce*? Also, could these processes *themselves* have been different? Could the process that led to the formation of our planet earth just as easily have led to the formation of a planet with a completely different makeup, a makeup that might have very significant consequences for what kind of life (if any) the planet could then support? Might the Colorado River not have existed at all, or might it not have led to the formation of the Grand Canyon? Might the Grand Canyon have been half its size, if causes and their effects had gone differently, thereby also affecting significantly the surrounding ecology and survival of local species? If the process of evolution had gone differently, might human beings not have existed at all? At first glance we might be tempted to say that, yes if the processes had gone differently, we could have had different outcomes in all of these areas, but a key question once again is: *could the processes have gone differently*? This question brings us to the concepts of chance and randomness, and related issues of necessity and determinism.

Chance, and the random nature of evolution

Before we begin to define our terminology in a more careful way, we need to look first at how it is claimed that some of these concepts work in evolutionary biology, according to leading thinkers in that field. The notions of chance and randomness are frequently appealed to by evolutionary biologists in their discussions of how evolution works; indeed these are important concepts particularly for those thinkers who are also secularists, and who wish to co-opt evolution to support their worldview. It would be fair to say that for these thinkers especially the notion of chance is vital in their discussion of evolution because they see it as a way of attacking or undermining the notion of design in the universe, and they wish to attack design as a way of arguing that there is no designer. These thinkers would include some leading writers on evolution over the past generation or so, including Dawkins, Coyne, Jacques Monod, and Ernst Mayr.[1]

But we must also acknowledge that there are other thinkers who believe that the evidence from a study of evolution itself, considered purely from a scientific point of view (and not co-opted as a way of supporting atheism) can at least be read as suggesting that there is a large element of chance and randomness involved in the process. One thinker in this camp is Stephen J. Gould, who proposed an interesting metaphor that became well known to illustrate this point: the notion of "replaying the tape of life." Gould proposes that we imagine the whole of evolutionary history as playing a tape, and suggests that if we rewound the tape, and played it a *second* time, we would end up with different species than the ones we have now, and almost certainly there would be no species of *Homo sapiens*. Gould calls this random aspect of evolutionary development "contingency," and, without clearly defining his terms, argues that the universe exhibits contingency, and not necessity. Gould puts the point this way:

> … any replay of the tape [of life] would lead evolution down a pathway radically different from the road actually taken. But the consequent differences in outcome do not imply that evolution is senseless, and without meaningful patterns; the divergent route of the replay would be just as interpretable, just as explainable *after* the fact, as the actual road. But the diversity of possible itineraries does demonstrate that eventual results cannot be predicted at the outset. Each step proceeds for cause, but no finale can be specified at the start, and none would ever occur a second time in the same way, because any pathway proceeds through thousands of improbable states. Alter any early event, ever so slightly and without apparent importance at the time, and evolution cascades into a radically different channel.[2]

The whole process, according to Gould, is contingent, which means that it could have gone differently to the way it actually did go. This is because there is a vast series of intermediate steps between the start of a causal process and its eventual ending in a recognizable species (or indeed in any event or happening in nature), and *any one of these steps could have gone differently*, and so the end result would be different. As an example, we might say that if a deadly virus had come along at an early stage in the evolution of our immediate ancestors, they might have been wiped out, and therefore we would not have come into existence. Or if a very bad winter had not come along 500,000 years ago and killed off a particular virus that was present in the atmosphere, we would have no modern elephants, and so forth. And Gould is suggesting that a deadly virus, or an extreme cold snap, could easily have occurred, and even though these events may appear insignificant in themselves at the time, their (cumulative) effect on evolutionary history is very significant indeed. So the existence of any particular species (or indeed event) is contingent, not necessary, a matter of chance, not design. If we begin our analysis by considering a species as an end result, he says, we could at least in theory retrace the steps that led to its coming into existence; however, if we look at the same process from its beginning onwards, we are *unable* to predict which species will emerge at the end, a point that again underlines the contingency of the species. So if we were to replay the tape of life all over again, Gould thinks that we would end up with a vastly different group of species than those we actually have. It is fair to say that Gould's understanding of the role of chance in evolution has been very influential in the discipline of evolutionary biology; it has become a part of the official theory, and an important part of how the theory is explained and taught.

Gould's argument is about the larger picture involving events in nature such as geological happenings, climatic changes, effects of predators on the environment and on species. Dawkins makes a similar type of argument, but with regard to mutations (though the underlying argument both are making is the same, even though the two approaches are often confused with each other). Gould thinks that there is a randomness or contingency in the larger picture of nature, and Dawkins mostly talks about randomness in terms of the mutations that occur in the development of the DNA of a particular species, and how these mutations affect the eventual structure of the species in question.

The word "mutation" means change; and it is a special term in biology to describe changes in DNA. It refers to a change in the DNA structure of the parents, which is then passed on to the offspring in the process of reproduction. These types of changes or mutations are caused by features of the universe such as the environment, chemical processes, diet, radiation, and by problems in the DNA itself (which themselves will have to have causes). Not all mutations that occur in cells are passed on to the offspring of the parent; only those mutations that occur in reproductive cells will be passed on, and so these are the ones that drive evolution. In short, some causal processes occur that change the structure of the DNA in a gene and this change can be beneficial to the organism, detrimental to it, or, what is more common, can have no effect on the organism's fitness to survive. According to the theory of evolution, new species came into existence, in the early stages of evolutionary history particularly, because genes mutated for various reasons, and gave rise over time to new characteristics, structures, and eventually to new life forms. The fittest of these life forms or species survived by means of the process of natural selection. This process took an incredibly long time, of course, but eventually it led to all of our present (complex) species.

In his discussion of randomness, Dawkins mostly discusses mutations.[3] He notes that when we are talking about the issue of species change, or changes within a species, we are really talking about mutations, which are sometimes described as occurring randomly (even by many biologists). He recognizes that we need to handle the concept of randomness carefully when talking about mutations, but argues that the existence of mutations (and therefore of species, as we have seen) is indeed random in one important sense, but not random in another sense. Dawkins acknowledges that mutations are non-random in the sense that they are *caused* by definite physical events; they can also be regarded as non-random because we can say, at least in some cases, which mutations are more likely to happen, and which are not more likely to happen (i.e., all genes are not equally likely to mutate, so the fact that some outcomes are broadly predictable means that the process is not random, a crucial point, to which we will return later). This means that we need to be careful when we say that mutations are "random"; to describe them as random sounds as if we mean to say that the mutation does not have a cause, that it somehow "just happens." Also, that it could just as easily not have happened. Of course, this can't be true. Mutations, like every event in the physical universe, have causes, and indeed part of scientific research includes trying to discover the causes, especially of mutations that lead to disease (e.g., those that cause sickle cell anemia, an inherited blood disorder), and then to prevent the causes, so that the mutations don't occur. So Dawkins is correct to recognize that when we describe mutations as "random," we don't mean to say that mutations have no causes. So we need to be more precise then about what we do mean.

He offers what he thinks is a more accurate understanding of what it means to describe a mutation as random. He says it means that no mutations occur in "anticipation of what would make life better for the animal."[4] This means that mutations occur with no regard to how they will affect an organism; some will affect it for the better and some for the worse, and some will have no effect; it is a matter of chance which mutations actually occur, as well as how they affect the organism. Natural selection works with whatever mutations occur, and so it is a matter of luck or chance which features of an organism end up conferring an *advantage* on it, and also, of course, a matter of luck whether these advantages actually help the organism to become the "fittest" in its particular environment. In short, those species that survive do so largely because of random mutations (in Dawkins's sense of the term), along with the process of natural selection; there is no design involved, no initial causal forces in advance driving one particular result rather than another. On this point, he is in fundamental agreement with Gould.

Elliot Sober defines the various random changes in mutations, which lead to what is called random variation in varieties and in species, as the fact that there is no known physical mechanism *that detects which mutations would be beneficial to an organism*, and that causes these mutations to occur.[5] Sober's definition is an attempt to capture his belief that when a mutation occurs in DNA, it occurs by chance, or perhaps more accurately, as we will see below, that its effect upon the organism is a matter of chance. It might be beneficial to it but it could just as easily have been detrimental or neutral with regard to an organism's development, or potential for survival.

Another very influential thinker on evolution, Jacques Monod, is very clear in his belief about the random nature of mutations:

> We call these events accidental; we say that they are random occurrences. And since they constitute the *only* possible source of modifications in the genetic text, itself the *sole* repository of the organism's hereditary structures, it necessarily follows that chance alone is at the source of every innovation, of all creation in the biosphere. Pure chance, absolutely free but blind, at the very root of the stupendous edifice of evolution: this central concept of modern biology is no longer one among other possible or even conceivable hypotheses. It is today the sole conceivable hypothesis, the only one that squares with observed and tested fact. And nothing warrants the supposition—or the hope—that on this score our position is likely ever to be revised.[6]

Monod is quite dogmatic in his statement of this position. He is sure that there is only one explanation for these mutations—chance—and that this is now a fact. According to Monod, this means that all of those mutations that occurred in evolutionary history that led eventually to our present complex species were a matter of luck. We might have had many other types of mutations, and they might have led to the emergence of different species, to fewer or no species at all. There might have been chaotic structural arrangements, haphazard species, little or no complexity, no fascinating biological organization, and so forth. Of course, in actual fact, mutations have led to enormously complicated life forms, but this is just a matter of chance, according to these evolutionary thinkers.

Monod presents his view on this matter with absolute certainty, and holds that it is not open to revision. Leaving aside that this kind of dogmatism has no place in science, especially with regard to the very complex notions of chance and necessity, and of their application in biology, notions that have been subject to a number of very different interpretations by many important thinkers, one can see why Monod wants this conclusion so badly. This is because he thinks that it gives him a good reason to then conclude that the emergence of any species, including *Homo sapiens*, is entirely a chance event, and not the result of a design plan. In this way, the presence of chance operating at the heart of biology can be presented as an argument against God's existence, as an argument against the compatibility of evolution and religion. Now whether he interprets evolution as operating by chance because he is an atheist already, and is looking to co-opt (a certain interpretation of) evolution to support his atheism (which would be an unacceptable biased reading of the evidence), or whether he is neutral and perhaps open on the question of a designer in nature, but honestly believes that the evidence supports the conclusion of a significant element of chance in the process of evolution and that this counts significantly against design, only he can judge.

But I suspect that many writing on this subject fit into the former category. However, it is our job to look at the theory of evolution dispassionately—in a neutral way—especially in the case we are interested in here with regard to mutations and the role of chance. Gould, Dawkins, Sober, and Monod, and other influential thinkers on various aspects of the relationship between evolution and religion, have all placed a large emphasis on the concepts of chance and randomness in evolution, and so we must now try to explain these concepts more clearly.

What does chance mean?

Our first task in addressing this complicated subject is to try to define our terms more clearly. In most discussions of these concepts, especially as they relate to the theory of evolution, very little attention is paid to the definition of terms, and then to the specifics of how the process of evolutionary change would work, given various interpretations of key concepts. As a result, the discussion is riddled with imprecision, vagueness, and confusion. So it is vital to be clear at the beginning about what we mean when we are talking about the process of evolution containing a large element of chance.

So let us begin by taking the term "chance," and looking at some popular ways the term is used.[7] One meaning of the term "chance" is to indicate that an event or a happening in nature or in our experience could have gone otherwise than it in fact did. So we might say that it was just a lucky chance that the tornado missed the building in which the occupants were taking shelter, meaning that the tornado, which swept across the town, could just as easily have hit the building. Another example might be that John was lucky the drive belt broke on his car today, and not yesterday when he was on his way to an important meeting. We think he was just lucky that it did not break yesterday, an event that would have caused him to miss the meeting. A second use of the concept of "chance" may be found in speaking of "games of chance," such as cards, slot machines, roulette, and perhaps in sports like horse racing. So we might say that it was a lucky chance that Ciaran won the jackpot when playing the slot machine because the odds or chances of getting the winning numbers are so low. Our speaking of chance here can be itself understood in two ways. The first meaning is that it was just a chance event that from all the possible combinations of numbers that could have come up on the machine, the jackpot numbers *just happened* to come up. The odds of the jackpot numbers actually coming up on a slot machine are very low indeed. The second meaning is that we might be emphasizing the point that it was just chance that *Ciaran* happened to be playing the machine when the winning combination came up. It could so easily have been someone else. Here we may be confusing chance with predictability, which I will say more about in a moment.

Applied to evolution and specifically to the phenomenon of mutations, we can note two areas where chance, understood especially in the first sense above, might occur. We might say that the mutation that led to the development of a specific trait that turned out to have adaptive value for the organism (i.e., that helped it to survive in its specific environment) happened by chance; it did not have to happen, and as Gould notes, if we played that part of natural history (that series of causal sequences) again, it would not happen. However, we must be very careful to note, again staying with Gould's particular take on it, that it is not just the mutation that occurred by chance, but also the whole environment that the organism (with its DNA subject to various mutations) finds itself in. This environment is also significantly affected by chance occurrences; indeed all of its features, like terrain, rocks, vegetation, climate, atmosphere, even location, are all to some extent chance events in the history of physics (using the concept of chance the way Gould is using it). This means that, for example, if the environment is particularly wet and this moisture plays an important part in the survival of local species, that the presence of the moisture just happened by chance. It may have come about because there was an avalanche in a nearby mountain range thereby allowing the flooding of the valley; but the avalanche might not have happened, and so forth. This is the way Gould and many others are thinking about the operation of causes in nature; those causes that did in fact happen did *not* have to happen, thereby introducing a significant element of chance into the occurrence of events.

Yet matters are not quite that simple. To see why, let us ask again what it means to say that a mutation occurs by chance. It means that the mutation did not have to occur, or could have been different in its structure. When we say that the mutation did not have to occur, do we mean that the mutation *has no cause*? It might sound as if this is what we mean, but a moment's reflection shows immediately that we cannot mean that the mutation has no cause because we know from basic logic and our knowledge of science that not only the mutation, but every ordinary physical event, has a cause (and most of our work in science is an attempt to discover these causes and perhaps to influence them). For example, if the mutation involves a change in the DNA that produces skin cells, we can say that it was caused by exposure to a certain chemical in one's diet. This is clear, and it would not be logical to suggest that the mutation had no cause in this sense (and so, for example, we can lessen the risk of developing skin cancer perhaps by changing our diets). This means that when an organism has a gene, and the gene mutates in one way rather than another, there is always a cause for the mutation (whether or not we can discover the cause). Of course this important but sometimes overlooked point is also true when applied to any other event in the universe as well, whether it is in the domain of biology or not. For instance, when we say that the Grand Canyon is an accident of geological development over many millions of years, do we mean then that the Grand Canyon has no cause? No; of course, we are fully aware that the formation of the Grand Canyon had a cause.

Another way of understanding the claim that a mutation occurs by chance is to say that, although it has a cause, the cause *might have been different* than it in fact was. This perhaps is a more obvious interpretation of what it means to say that an event, such as a mutation, occurred by chance. Suppose there is a plant growing along a river bed that secretes over time a chemical into the water; suppose further that a beaver drinks from the chemical-tainted river, and that the chemical alters the DNA of the beaver, over time. It then passes on this mutated DNA to its offspring, and let's say that the effect is considerable because it causes the future offspring to become blind, and eventually this particular species of beaver is wiped out by its predators. Now to say that this event occurred by chance is to say that a number of *other possible events might have happened instead*: that the beaver might not have drunk from the river, in which case the chemical would not have been able to affect his DNA; that a different type of plant might have been growing along the river bed; that the same type of plant might be growing but not have the chemical present; that the plant might have been growing too far from the river for the chemical to enter the water, and so forth. It might be that this particular plant is only growing in the river bed because a previous huge storm removed the tree cover thereby allowing these plants to receive the light they need to grow and thrive, but *the storm might not have happened*. And even if it had happened, the seeds from the plants might not have landed just where they needed to in order to be nourished by the river water and soil nutrients, and so forth. It might be that this particular plant needs a certain type of soil nutrient to survive and that the nutrient is only found in this one place near the river, and so on. The argument seems to be that because any one of these events—and the thousands of other events that go into making up the total causal environment (including also, of course, all of those involved in the beaver's biological structure)—might not have happened, or could have been different than they in fact were, that it is only by chance that the beaver's DNA was detrimentally affected. Of course, the reverse is true as well, it is only by chance that *favorable* mutations occur, and more generally, that any organism or species is able to survive eventually throughout the whole process.

Sometimes we say that events occur by "pure chance" (as in the Monod quote above), a phrase that perhaps is meant to convey two points, again that are not always clearly distinguished. The first is that there are so many events that have to occur *together* in order for a species to survive at all, let alone to become complex, that we can say that it was incredible that any did survive, and became very complex. It looks more likely that given the chance nature of *each step* that the final result (for example, a species of elephant) would never be able to develop and survive at all, perhaps not even in the case of very simple life forms. And yet we now have an incredible array of extraordinarily complex species. The second point being conveyed is that there is no one directing the process, that the final outcome is not intended or designed (despite appearances to the contrary). We will return to this crucial matter in just a moment.

In the language of necessity and contingency, we can say that there is nothing *necessary* about the existence of the plants that secrete the chemicals into the river, they *just happened* to exist there because the river *just happened* to break through from the mountain above and this chance event facilitated plant life. The plant seed *just happened* to land near the river, and the sunlight needed for the growth of the plant *just happened* to break through because the tree cover *just happened* to be destroyed by a storm, and so forth. (And let us not forget that the existence of the sun, the earth, and trees, etc. is also all a matter of chance on this view.) This explanation of what chance means is the reason why many people look at what has emerged from the process of evolution, and believe that the claim that it all happened by chance is unbelievable, and very difficult *logically* to accept. It seems irrational to look at the complexity of life, especially at our level of *Homo sapiens*, and to argue that this just all happened to come about because all of the millions of causes and effects, including those involved in the process of mutation, that are required to bring about a complex species *just happened to come about*. And that the same thing had to happen with the environment as well, not just the local one, but also the planetary one, and the galaxy one. This is why many reasonable people look at the theory of evolution and find it hard to believe that it is not directed or designed in some way, because the evidence from the theory itself—the complexity of life and its environment—suggests otherwise.

The meaning of randomness

We must now turn to the notion of randomness. Our first task is to distinguish between the concepts of chance and randomness. We noted above that one understanding of the term "pure chance" in evolution is the claim that the process is not designed or directed toward particular goals. So when a mutation, for example, happens that proves favorable for an organism, this was not caused to happen so that it would be favorable for the organism; it just happened by chance, and it could just as easily not have happened, could just as easily have gone differently, in the way we discussed above. This meaning of pure chance helps us to bring out the difference between the meanings of the terms chance and randomness, and also to see how nevertheless in the physical world *they are inextricably related*. The distinction between the notions of "chance" and "randomness" will go a long way to helping us think more clearly about how to understand the topic of chance in nature.

We noted above that Dawkins is aware that there is a danger of confusion when we say that mutations are "random." He recognizes that living organisms along with species give the appearance of being designed, but then he needs some way to explain why this is just an appearance and not a reality. It is precisely in answering this question that we need to focus on the concept of randomness. Dawkins acknowledges that to say that a mutation occurs by chance does *not* mean that there is no cause for the mutation. Of course, Dawkins has an agenda in the sense that he wants the evidence to show that all living organisms and all species came about by chance. He knows this claim looks far-fetched in the extreme, so he turns to randomness to develop the argument. As noted above, he defines "randomness" as the fact that no mutations occur in anticipation of what would make life better for the organism. This is a good way to think about randomness in biology, and a good way to think about what biologists often mean, as in the above view of Monod for instance, when they say that mutations are random (even though sometimes they use the word chance instead of random, and indeed sometimes use the terms interchangeably). But we must keep them distinct from now on because they play different roles in the argument, even though they are inextricably linked. So to keep them distinct we need to offer more precise definitions to capture more accurately their different meanings.

To say that an event came about by chance means that the prior cause(s) of the event could have been otherwise, and so the event might not have happened (in the way explained above). *So randomness then means that there is no goal or purpose or end toward which a chain of causes and effects is striving (for example, in the development of an organism), and which the various causes (for example, in mutations) are aimed at bringing about, or to which they are contributing*. Randomness therefore relates to the question of teleology. The concept of teleology goes back to Aristotle and has been discussed by many thinkers since.[8] It refers to the view that nature has goals and purposes built into it, which it is trying to bring about, or toward which it is striving, as

Aristotle put it. To say that mutations occur randomly is to deny that there is any teleology in nature, to deny that nature has teleological goals. When various thinkers say that mutations are random, or that the evolutionary process is random, they mean that the process leading to mutations has no goal in mind, that it just happens; they also mean that there is nothing, as Sober put it, in the physical process of mutation that *directs* the mutation toward a favorable end or goal for the organism; and then they also usually add that there is nothing outside of the organism either that is directing the process of mutation. So this is another way of understanding why the theory of evolution is controversial, because many understand it as a critique of teleology in nature. Whether it is or not is not quite clear yet, but it is clear that many like Dawkins press it into service in this way. And Darwin himself of course knew that even if he had no intention of discussing teleology, and even if teleology has no real place within biology, that his theory did appear, on the surface at least, to have some significance for teleology (as a critique of Paley, as we noted in Chapter 2), and that this could make it a controversial theory.

So we now have clear, more precise definitions of chance and randomness. In this discussion it is best to keep these respective meanings of the terms clear, rather than using the terms interchangeably. *To say that an event occurred by chance is to say that its prior cause(s) might have been different and so the event might not have happened. To say that an event is random means that it occurs without regard to any goal or purpose or end result.* The event is not aimed at bringing about any particular outcome, and the events that contribute to it are not themselves trying to bring about any particular outcome either.

Now let us examine why these two concepts are nevertheless inextricably linked. *The reason that chance and randomness are inextricably linked is because the cause(s) of an event, and by extension the causes of a chain of events, determine the outcome or the goal of the chain.* Let us illustrate this point with a few simple examples, and then we will apply it on a larger scale to physics and the universe, and then (in the next chapter) we will apply it in more detail to biology and evolution. We will consider a case of human causation first because this is helpful in allowing us to see what it means to have a process that is directed toward an outcome that is planned or intended (and so is not random). Suppose I am building a shed. First, I cut out the foundation from sheets of wood; next I cut out the sides of the shed from the wood, then the roof. I nail the sides to the foundation, attach the roof, and leave a space on one side for a door, and on the other side, I cut out space in the sheet of wood for a window. In this example, the sides of the shed did not come about by accident. Two things happened to bring them about: the first is that I had a plan of the shed in my mind; the second is that I then used saws and chisels to work the wood into the proper shape. I am only interested for the moment with the second point. Given the way I used the saws and chisels on the wood, the outcome could not have been otherwise. Given that I had a large rectangular piece of wood to begin with, and that I used a saw to cut it into a square of a certain size, it could not have been otherwise than it is. This means that it could not now be a circle, or be cut into three strips, or cut into rails. *Given the way I used the saw on the wood* (the *causal effects* of the saw on the wood), it has to be the way that it is. This means that the final outcome of the sheet of wood in this case is not random. *There was a certain way it had to turn out given the causes that affected it.* And this is another way to express the point that chance and randomness are inextricably linked. The cause, in fact, determines the effect, or to put it another way, the cause determines the end result (or goal). Given just *that* cause, no other outcome is possible. If the causes were different, then the outcome could be different. If I had cut the wood differently, say by using a router instead of a saw, there would be different effects and different outcomes.

So there are three points to take away from this example, and we will see them again in further examples that take us closer to the issues in which we are interested. The first point is that given the cause, the effect has to occur, and is not random. The second point is that *if* the cause that produced the effect could have been different (could just as easily have been some other cause), then the outcome does not have to come about, and therefore perhaps could be said to be random. The third point is that an outside observer looking at an event or an effect, after the fact as it were, could try to work out whether the event was random or not *based on the likelihood that it could have been otherwise than it in fact is.* The observer could reason backwards as it were from the event to its likely causes, and could then ask if the causes could have been otherwise than they in fact were. (Note that when we look at the shed {the effects} and reason backwards to their causes, we are not thinking about the agent who brought about the causes, but about the causes themselves, the saws, chisels, hammers, and so forth that were used to construct the shed.) This example shows that chance and randomness, while different concepts, are inextricably linked in any chain of causes, any causal process, in the coming about of any event. This is because if the cause that brings about the event could not have been otherwise, *then the event does not occur randomly.* But if the cause could have been otherwise, then the event can be said to be random. (Recall that "random" means that the event is not aimed at bringing about a certain goal or end product from the causal process, and "chance" means that the cause of the event could have been otherwise.)

One might be tempted to think that this line of argument only applies to events that are brought about by an intelligent agent, as in our example of building the shed. It might be thought that if there is no agent involved, we would have to look at the event differently and accept that the cause could have been otherwise, and therefore that the final outcome of the causal chain(s) is random. And this line of reasoning might then apply in both physics and biology. However, in the next section I will argue that this is not the case with regard to physics, and in the next chapter I will apply my general argument to the process of evolution itself.

Determinism and chance in the universe

As we noted earlier, Sober has attempted to capture the relationship between chance and randomness by defining randomness as there being no known physical mechanism that detects which mutations would be beneficial to an organism, and that causes these mutations to occur. Here Sober is introducing a third factor into the discussion in addition to chance and randomness; this third factor is a physical mechanism that would somehow "direct" the mutations toward a certain outcome. Sober thinks there is no such mechanism, and so therefore mutations must be random, but he admits that we cannot rule out that a non-physical mechanism, such as God, might be guiding the mutations from outside the process, as it were (a point we will come back to in a later chapter). However, there is no need to appeal to God yet in our argument, for the essential point is that, despite what Sober says, we can show that nature is moving toward goals without needing to introduce a new physical mechanism into nature. We simply need to appreciate a point that we know is true but often forget when we get bogged down in the thicket of chance and necessity. *This is the simple fact that it is not true that the cause (or more accurately, causes) of any event in the universe could have been otherwise that they in fact were. And if a cause could not have been otherwise, this means that the end result of a causal chain could not have been otherwise either.* In short, we are looking at a completely deterministic universe in the realm of the laws of physics and science (leaving out for the moment human free will, and any action by God in the world). And once we come to realize and focus on this point, it changes the discussion of chance and evolution completely. Sober misses the mark because he fails to realize that *the causal chains themselves are physical mechanisms.*

As noted at the beginning of this chapter, philosophers have traditionally been very interested in the question as to whether there is any genuine chance in the universe (leaving out as noted human free will, or intervention by God in creation). They want to know what about the universe has to be the way it is, and could not be otherwise. We noted above the areas of nature and experience that might give rise to this question, but for now, let us concentrate on asking it about matters in physics. Keeping the subject on a very practical level, we are interested in questions like: could the Rocky Mountains have been different than they are? Could they have been much higher, or might they have extended south only to Wyoming instead of New Mexico? Similar questions can be asked about *any other feature* of the universe: did the Ice Age have to happen, or might it not have occurred at all? Was its occurrence a matter of chance? Did the earth have to form? Did the galaxy of Andromeda have to come into existence, or might there have been no such galaxy? Did the atoms that formed out of the big bang (in the first 400,000 years) have to form? Might they have formed differently, so that the fundamental structure of the universe therefore would be different that it in fact is? In thinking about such matters we must consider the laws of physics, and we must ask again the key question we identified earlier: *could things have gone differently*?

The first point to consider here is the laws of physics. Our work in the various sciences shows us that there are "regularities" that occur in nature; these are empirically detectable patterns in how nature behaves. We discover these as we practice science, because we discover that nature behaves in uniform ways most of the time, and so these "regularities" have been codified and established as the laws of science. These include the laws of physics and chemistry, perhaps the two most basic sciences (along with mathematics) for understanding the physical universe. Examples of the laws of physics would include Newton's law of gravitation; Kepler's laws of planetary motion (his first law states that the planets circle around the Sun in elliptical orbits, with the Sun at one focus of the ellipse); Ohm's law relating to electrical circuits; laws of chemistry based around the elements in the periodic table; and many others, both simple and complex. A simple "law" from the point of view of our ordinary experience might be that an object with a greater mass will crush an object of lesser mass if it lands on top of it, all other things being equal (this "regularity" involves several laws of physics). Or that an object with a lighter color will be harder to see against a lighter background than a darker object. Or that water will weaken the composition of water pipes over time (and so periodically the pipes will need to be repaired). These are ordinary everyday *causal facts* that we are familiar with in our experience, and, even though most of us may not often think of them as being governed by or explained by the "laws of science," their existence is remarkable from a philosophical point of view because they indicate that there is an underlying order in the universe.

These laws have been progressively discovered, codified and then utilized by the various sciences over the centuries. Indeed, let us not forget that *they make science possible*, for it is only because causes and effects in the universe follow laws, or fall into patterns of regularity, that we can formulate explanations of events, and make predictions about what will happen with regard to future events. If there were no laws, there would be no science; in this sense, the laws are prior to science, and are not explained by science; rather scientists discover them and then utilize them in doing science. Of course, it is a quite remarkable fact that the universe follows laws like these, regularities and consistencies in nature that determine how nature operates. So we need to bring out more clearly the connection between the laws of physics that operate in our universe, and the nature of causation, and its relation to randomness. Once we see the correct relationship between these three concepts, we will realize that there is no chance operating in the universe.

We need to state more precisely our central question about chance and randomness in the universe. That question is: is there anything that happens in nature, or indeed in the physical universe as a whole, that is not governed by the laws of physics? The answer to this question in modern physics *at the practical level of doing science* is generally no, especially at the atomic level (we will come back to the subatomic level, and to the theoretical level more generally, in the next chapter). What this means is that whenever anything happens in the universe, there is a cause for it, and the cause operates according to, or obeys, or can be explained in terms of, the laws of science. It would be difficult to do science if we believed that events did not obey physical laws. But this means that there is an underlying *determinism* in how nature behaves. This conclusion has enormous significance. Before explaining in more detail what it means to say that there is an underlying determinism in the universe, let us illustrate further by means of a few examples.

We will turn first to the development of the physical universe; this includes the development of the environment that emerges on various planets in the universe, such as earth, and so will also lead eventually to the development of our natural world here on earth. Our natural world will also then include the environment in which later organisms and then species came into existence; this environment has a very significant impact on the process of evolution. We can tell this story generally as follows, based on current scientific evidence. The universe began between 10 and 20 billion years ago from a huge explosion called the big bang. There was nothing before the explosion; it was the first event in the universe. There was no matter and energy or laws of physics before the big bang; rather, all of these features came into existence at the big bang. There was no space or time before the big bang; these too came into existence at the big bang. Then out of the effects of the explosion there came about over time the various ingredients that make up the stuff of our universe, such as protons, quarks, and eventually atoms, and later even more complex molecular structures.[9] These initial states—which are the subject of much speculation and argument—then determined the next states of the universe *completely,* right up to the formation of galaxies (our solar system is thought to be 5 billion years old, and our neighboring galaxy, Andromeda, is thought to be 8–10 billon years old), and eventually to the formation of the earth (about 4.5 billion years ago).

A more specific example will help to illustrate this process. Although we don't know exactly how the Grand Canyon formed on earth, we can identify several events that contributed to its formation, including the Colorado River, the activity of volcanoes, the Ice Age, the presence of various gases in earth's atmosphere, and so forth. These events were part of a causal chain the end result of which was the Grand Canyon. Within each event the same causal analysis applies; there is a chain of causes that produced that event as an end result (for instance, there is a detailed causal history leading to the occurrence of the Ice Age). One might note that the environment also has an influence; for example, if the gases are forming in a specific environment, then the environment will have a causal influence on the gases. This is perfectly true, but the environment came about in exactly the same way, as a chain of causes, each cause in the chain leading to the next event, which leads to the next event, and so on to the *end result* (or *"goal"*). We all know that this is how physical causation works in the universe, but when we talk about the abstract concepts of chance and randomness, we tend to forget this point.

I would like to use the analogy of car engine to remind ourselves of how physics works. Suppose your car will not steer properly, and the mechanic looks for the cause. He explains to you what happened. The engine had developed an oil leak from the valve cover; the oil had fallen over time from the top of the engine down onto the power steering pump belt; the oil weakened the belt, which eventually snapped; without the belt, the power steering pump pulley would not turn, and so the pump would not operate, and so you are unable to steer the car properly. None of these events occurred by chance; they are all part of a causal chain, each leading to the next causal effect. Not for a moment would the mechanic think that an event happened in the engine by chance, say that there is no cause for the fact that the pump pulley is not turning, or that, *given the prior causal conditions* in the engine, the cause of the belt snapping could have been different. It is not even remotely possible that the mechanic would consider this; nor would he return your car to you with the observation that "your car does not obey the laws of physics!" *It is the same with nature.*

The difficulty we have with regard to cause and effect in the universe is that we cannot reconstruct the causal story about how things happened in the past because the subject matter is simply too vast and too complicated for us to be able to explicate its full causal history. We cannot reconstruct the full causal history of the existence of the Amazon river, for instance, but we do not doubt that in an account of its full causal history each effect will require *its specific prior cause*, right back to the beginning of the causal chain. Not for a moment can we entertain the view that one piece of the causal chain, say the Ice Age carving out the river bed, could not have happened *given its specific prior* cause (a specific prior cause that had *itself* to happen, *given its specific prior cause*, and so forth).

We can use another analogy of the balls on a pool table to illustrate this point. When one "breaks" the pool balls around the table with the cue ball, the location each ball ends up in is not a matter of chance, though we often speak loosely as if it is. But what we really mean is that we can't *predict* where the balls will go, but not that where they go is unpredictable in itself, that is, happens by chance. Where each ball goes is precisely *determined* by factors such as its speed and weight, the speed and angle that the other balls hit it, by the size, shape, and structure of the table, the environment of the room. Ball A (given its characteristics) has to end up in a certain precise location *given the cause of its movement* (being hit by Ball B with *its* characteristics on this *particular* table, etc.). The cause of A's movement can only be changed if B is changed, and B can only be changed if the cause of its movement is changed, and so forth. (This is why it is possible to construct computer models to play pool accurately and fairly—meaning that if in computer pool you hit the ball at the wrong angle it will not go into the pocket!) It is the same in the universe.

What are the implications of these points for determinism and chance? The first very important implication is that we live in a *deterministic universe*. Determinism may be understood as the view that every present state of the universe is caused directly (or is completely determined) by prior states together with the laws of physics (I am not advocating a complete determinism because I am excluding free human actions and any intervention by God in creation).[10] This thesis also means that the present state of events, together with the laws of physics, determines every future event (again, excluding intervention—causation by an agent—from either man or God). It is a thesis that has been much discussed by philosophers and scientists, and the French astronomer, Pierre Laplace, was one of the first to seriously propose (in 1814) the theory as the best way to understand how our universe works.[11]

Laplace also noted another point about determinism, as least as it applies to the physical universe: that if we knew every single ingredient and everything about its behavior in any physical system governed by the laws of physics, we could predict with certainty the future states of that system. To say that the universe is governed by determinism is to say that the formation of the Grand Canyon is completely brought about by its chain of prior causes. It is to say that the current state of the drive belt in your car is determined precisely and completely by the structure of the belt when it was first put on the car (which is itself determined by its ingredients, manufacturing process, etc.) together with the state of the other parts of the engine, and the belt's interaction with those parts, such as the composition and speed of the pulleys it travels on, the quantity and composition of the oil dripping on the belt, and the heat emanating from the engine block. The same is true for the formation of the earth given the causal changes out of which it emerged. So one implication is this: we live in a completely deterministic physical universe.

A second implication is that there are no chance events in this universe. This means that nothing occurs without a cause, of course, but it also means that *the causes that bring about events could not have been otherwise, and so the events had to happen*. Determinism changes our understanding of causal happenings in nature because if the causes could not have been otherwise, then the end results must come about, and this has great implications for the question of teleology. This is illustrated clearly in our various examples above. The Grand Canyon had to form in the way it did given its prior causes, and those causes had to be the way they were, given their *prior* causes, and so on back to the beginning of the causal chain. If one of the earlier causes had been different then the outcome would be different, but one of the earlier causes *could not have been different* because its prior cause would then have to be different, but this could not be different unless its prior cause could have been different (given its prior cause), and so forth.

It might be objected that the fact that specific events may have more than one cause could be offered as an argument against my position here. The objection is based on the fact that a particular event does not have to have the cause it actually had for it could have been brought about by any one of a number of causes. So, for instance, it might be argued that if a person becomes infected with influenza, that any one of four prior causes might have been responsible for bringing about the flu. Of course this is true, but it does not follow from it that the actual cause did not have to bring about the actual event (the flu). My point is that given one particular prior cause, then the event that follows *had to happen*. It is also true that the actual prior cause itself has to happen, even if it is the case that more than one cause might lead to the same event. So it could be true, to take another example, that a strong wind, or heavy rain, or a volcano erupting, could cause a boulder to fall from the top of a mountain. But if the boulder actually falls, there must be an *actual cause* of it falling, and given this cause, the boulder must fall. If we find that a heavy rain had occurred and had weakened the soil around the boulder, then on the view I have been developing here this had to be the cause of the boulder falling. This is because the rain had to fall when it did (given its prior cause), and the other causes could not have happened because their prior causes did not happen, and so forth.

Sometimes it is objected that what I have said so far is true as far as it goes, but that it does not take into account multiple causes coming together to produce an event, or an outcome. The argument is that just as the culmination of casual chain A is about to interact with causal chain B, some other cause could interfere with A, preventing it from interacting with B, and so it is at the level of causal chains interacting with each other that one might find an element of chance in the process.[12] It should be clear that this objection does not take full account of the fact that each effect in a causal chain is itself determined by prior causes, and so the whole causal chain must follow the path it does follow. But this is also true of the interaction of one causal chain with another; the interaction must happen, given the prior causes, and so the outcome would still be determined. The fact that we cannot predict, and likely will never know, the types of interactions that occur throughout the universe and throughout nature, does not alter the conclusion that the events that occurred had to occur, given the various causal chains that led to them.

This means that if a boulder falls off a cliff, we can argue that it would not have fallen if any one of seven prior events had not happened (erosion of the large rock face that produced boulders; weakening of soil though rain; soil drying out due to sunlight; removal of tree vegetation due to disease; animals trampling the earth around the rocks, etc.), but the fact is that all seven prior events had to happen, given their prior causes (which themselves had to happen given their prior causes), and so the boulder then had to fall from the mountain. Suppose we notice that all the trees have been destroyed by disease, except one, which is holding the boulder in place; otherwise it would fall. It is still true that there is a causal story for why the other trees died of disease and the remaining one did not, and we can also easily explain why it is holding the boulder in place by appeal to the laws of physics. There is no chance at all involved in any of these events and their causal histories; no event that does not have a prior cause, no effect that might not have happened given *its* prior cause. Therefore, the end result *must* happen given the prior causal chains, and so nothing happens randomly. This is how physics (and the scientific method) works.

A physicist once offered the following as an objection to the view I have been developing. He argued that exposure to sun rays can cause skin cancer in humans. The process involves the skin (say on a person's arm) being exposed to the light of the sun over a period of time. Some of the skin cells will develop cancer, and some will not. The physicist argued that we know in general that some skin cells will develop cancer in this way, but not *which* ones, and the cells that develop it do so randomly. My argument is that nothing random like this happens in physics because the actual cells that develop the cancer must do so for some causal reasons (say they have weaker membranes than the other cells). There must be, logically, a cause for why one cell develops skin cancer and one does not, and this means that given this cause, and all other conditions being equal (which they *must* be, given their *prior* causes), this cell will develop cancer. There is also a causal reason for why this particular cell has a weaker membrane than other cells. Not for a moment can we believe that the weaker membrane, or the cancer developing in a particular cell, might have no cause, nor can we accept that given *its* prior causes, the present cause of the cancer could then have been different (it could only have been different if the prior causes had been different, but these could only have been different if *their* prior causes had been different). This is why one skin cell develops cancer, and another does not. Of course, we may not be able to *predict* from a group of skin cells which ones would develop cancer and which would not because there are too many (unknown or hidden) variables involved, and we cannot have precise knowledge of the specific causal history, including the environmental history, of all of them. But we must not confuse our inability to predict an event with the claim that the event has no cause, or that some other event could have happened in its place. (I will return to the difference between chance and randomness, and predictability and probability, in the next chapter, in our discussion of these concepts as they apply to the processes of evolution.)

It is also essential to point out that this is also true of the *interaction* of causal chains such as one would find in the formation of the Grand Canyon, such as the coming together of various gases in a specific type of atmosphere that would later contribute to the formation of glaciers on the earth's surface. Given the starting conditions, there is nothing that can go differently. This means that the end result is completely determined unless the causal chains are interrupted in some way and they can't be in our universe (without the intervention of human free will or God). One can't say, for instance, that the heat that was influential in the formation of plants in a forest might have been interrupted by rain! This is because for this to happen the causal conditions that produced the heat would have to be different (to produce the rain), and these can't be different unless *their* prior causal conditions were changed, and these can't be changed, unless *their* prior conditions were changed, and so forth. There is no coming together of causal chains by coincidence, on this deterministic view. So we are led back then to consider the *beginning ingredients* out of which the initial causal chains emerge, and how they got that way. The fact of the existence of this beginning, and its makeup, and what it leads to now takes on an enormous significance. This is like saying in the pool table example that the end result of the balls is determined by the initial state of the table at the beginning of the causal chain, which raises the key question of how the table and its content got its initial properties; that is to ask, *who set it up and why*?

In this chapter, we have explained the role claimed for the phenomenon of chance in the universe by various thinkers, and then have gone on to examine how the domain of physics and the study of the physical universe actually works. We have seen that it is hard to avoid a conclusion of determinism about the way that nature operates. I suggest that this underlying determinism is presupposed by physics, and that this is how science works. When scientists look for the cause of an effect, they assume a determinism about the workings of nature; they assume that if A was the cause, then B had to be the effect, or that if B is the effect, then A is the cause; that if C occurs, D will occur, and that if F does not occur, C and D could not have been present; that for G to have occurred, A, B, and F must have been the causes, and so forth. Working scientists assume that nothing happens by chance; that effects have causes, and that given such and such a cause(s), the end result *must* come about. This is what enables scientists to offer explanations for the occurrence of events, to make predictions about what will happen in the future, and to offer guidance about how to prevent events from occurring (such as disease or global warming). Perhaps in physics we are more aware of the deterministic implications of the laws of nature and the way they operate in science, but in biology we seem to forget them, and we talk as if causes do not have to lead to effects, and begin to describe events in biology as "random" in a way we would never do in physics or chemistry. Many of these issues of course are hotly disputed among various thinkers, and there are no agreed upon answers to most of the questions I have been asking in this chapter. My argument is that at least at the level of ordinary physics, nature operates in a deterministic fashion, and that this is the standard position adopted by working scientists (whatever philosophers and scientists might say about the theoretical status of this working position, a topic I will come back to in the next chapter in a brief discussion of quantum theory).

In our next chapter, we will apply our thinking on these matters specifically to biology and the theory of evolution, and will discuss further the notions of chance and randomness in evolution, and also distinguish between chance, and predictability and probability. We will also consider some objections that might be raised against the view that the universe is deterministic, especially objections based on quantum mechanics. All of this will be very enlightening for our discussion of the question as to whether any chance operates at the level of biology. It will also return us once again to the questions of the origin of the universe, and the specific structure of the universe, with its specific laws, all of which lead in a deterministic way to the emergence of the various complex species we find on earth, including *Homo sapiens*.

6

Chance and Randomness in Evolution

In the previous chapter, we introduced the manner in which evolutionary biologists use the concepts of chance and randomness in evolution and then went on to consider these concepts in some detail, primarily in the domain of physics. We now need to apply the arguments of that chapter—specifically, the argument that there is no chance at all operating in nature—to biology, and especially to evolution. If it is the case that the universe is deterministic, then this conclusion will affect all areas that involve causation, including biology. This is a conclusion we are inclined to recognize more easily when we are talking about the domain of physics, but somehow we do not seem to appreciate its significance in the domain of biology, even though it is just as applicable in that domain. I suspect that many leading biologists who are inclined to accept some form of determinism in physics, at least as a working hypothesis, forget that this determinism also applies to their work in biology. The universe is either operating in a deterministic fashion or it is not; if it is not, then elements of chance will be present in causal processes throughout the universe; if it is, then there are no elements of chance in any causal process. This chapter will attempt to bring out the implications of the latter view for the theory of evolution.

Evolution, determinism, and the absence of chance

We can first apply our conclusions concerning chance to the phenomenon of mutations in the DNA of organisms, a phenomenon we have already explained in Chapter 5. We noted that mutations or changes in DNA can occur at any level of an organism's development; these mutations then have an effect on development. The consequences of this effect can be for the detriment or betterment of the organism, or may be neutral with regard to the adaptability and survivability of the organism, all *given* its environmental conditions. A well-known example of a mutation that is detrimental to humans is the mutation in the gene responsible for hemoglobin production; an error in this gene often leads to sickle cell disease. From the point of view of our discussion, the key question is: does this mutation occur in this particular gene *by chance*? We know from our previous definition of chance that we are asking if the cause of this mutation could have been otherwise than it in fact was? The answer is no because whatever caused the mutation was itself caused, and *given* this cause, the mutation *had* to occur. Was there also a cause for this prior cause? Of course. We would not hesitate to give this answer if we were talking about the domain of physics, if, for instance, we were discussing the causal process that led to the formation of the Grand Canyon. It is the same with any causal process we are studying in biology.

Let us also not overlook the enormously significant point that we can apply the same reasoning to the causal sequences that produced the environment that organisms and species find themselves in. The environment has to be the way it is all over the earth, and indeed all over the universe, given the initial starting conditions. This is because each event in the causal chains is caused by the preceding step, and could not be otherwise, right back to the beginning of those chains. Moreover, intersecting causal chains *must* intersect, given their prior causes. Evolutionary biologists sometimes overlook the fact that on the view that mutations occur by chance, one will also have to say that the environment occurs completely by chance, because the same understanding of the nature of causation must apply in both biology and physics. However, if there is no chance operating in physics, as I have been arguing, then there is no chance operating in biology either.

What we have been drawing attention to is that there is *a complete causal story* for why an organism has the biological and anatomical structure it has (although we have little idea of the specifics of this story in particular cases, and never will have). There is also a complete causal story to explain the origin of the type of environment in which an organism lives. So when we speak of chance in evolution, we are speaking too loosely. We are often confusing our inability to describe how something happened, or to *predict* what will happen, with the view that when it happened it happened by chance (i.e., that its prior cause did not occur by necessity; that its prior cause could have been otherwise than it in fact was). What we should be saying is that we can't explain exactly how things happened, or predict what will happen, because we don't know the full causal story due to gaps in our knowledge, not because there is no full causal story, *and not because each step in the causal chain is contingent*. There are no contingent steps at all in the causal chains; no steps that could have been otherwise, given their prior causes. We overlook the fact that for *every effect* that occurs in biology, there is a specific cause of this effect, *including for every supposedly ("chance" or "random") mutation, and for every environmental change, right back to the beginning of time*! Evolutionary theorists sometimes forget this, or ignore it, when they are talking about the process of evolution.

Jerry Coyne has suggested that there is no known biological way to increase the probability that a mutation will meet the current adaptive needs of the organism, and he suggests that a mutation should be described as "indifferent" rather than random.[1] It is this reading of evolution that renders the coming into existence of organisms and species so unlikely, because one is then supposed to conclude that the emergence of incredibly complex organisms, together with the general progressive nature of evolution, just happened by chance, and that it could just as easily not have happened. The complexity of organisms and the progressive directionality of evolution give us pause that the underlying account of chance that is supposed to have produced these phenomena may be incorrect. To support his claim, Coyne offers the example that light coats would be more useful to mice living on sand dunes, but then suggests that their chances of getting a useful mutation that leads to light coats are no greater than mice developing the same mutation that live on dark soil.

Let us note a few things about this example before looking at how it would be understood in the light of my argument. One important point is that, according to Coyne's understanding of the way the process of evolution operates, the mice get the mutation or set of mutations that produces light coats by chance. Whatever caused the mice's DNA to change to give them light coats could have been otherwise—in the way we discussed in the previous chapter—and so the mice might have ended up with dark coats, or coats of some other color (or a bird might have had a different wing structure than the one it has, or no wing at all, to refer to an example given by Francisco Ayala[2]). Second, this type of reasoning applies to all of the mice's features, of course, something we often forget when thinking about these cases. Every feature the mice have in their biological and anatomical makeup—*every* feature—came about, either directly or indirectly, in the same way, due to a process of mutation that could have been otherwise. It is not just that the mice exist with everything already in place except their coats, which a mutation then produced, and this change just happened to help them blend in with the color of the sand dunes, thereby giving them an advantage that aided their survival. This is much too simple. The same thing happened when the mice developed their legs; indeed, not just their legs, but also the various complex parts of their legs, and indeed the structure of their feet and toes, and the nature of their leg bones, etc., and so on for *all* of their features. We must remember that this interpretation of evolution is claiming that all of these features came about in the same way as the mice's coats came about, and that they all just happened to be beneficial to the organism, that they all could have been very different, and the mice could *just as easily* not have survived. The third significant point is that we must also consider very carefully the role of the environment in the whole process. We have been looking at the issues up to now mainly from the side of the organism, how changes in its DNA bring about changes in its features, but we must not overlook the fact that the environment of the organism is often responsible for bringing about DNA changes. In addition, with various DNA changes the organism has to survive in an already existing environment that will obviously constrain by its structure to some extent what kind of organism can survive in it, and even whether it can survive. We will come back to these points in the exposition that follows, especially the third point.

How does my argument about chance and necessity apply to this type of argument proposed by Coyne, which is a very typical argument in evolutionary biology, one that is found in many expository works and textbooks, and one that seems to be now adopted as part of the official way to explain and teach evolution?[3] The argument is also often accompanied by a calculation of the probability of whether a particular species is likely to undergo a mutation of a certain sort (for example, whether a particular person is likely to get sickle cell anemia, given certain genetic presuppositions in the parents), and this is a subtle way of further emphasizing the fact that there is a large element of chance involved in the process of mutation, a process that can lead to significant changes in species over long periods of time. My argument means that we need to reread Coyne's example in a different way to the way he reads it. He interprets this example of the mice and their environment as an example of chance and randomness on both sides of evolution, the mutation side, and the environmental side (the process of natural selection). This is because he thinks that the mutation that resulted in a change to the mice DNA that led to light coats did not have to occur; that some other cause could have happened and produced a different effect in the mice's DNA. I am suggesting that Coyne is wrong about this, and has simply not paid sufficient attention to the nature of causation. Coyne also thinks that the mice could have been born into a different environment, one detrimental, which would have prevented them from adapting well and would have threatened their survival. He thinks it is a matter of chance that they evolved in an environment that just happened to be adaptive for them—so adaptive that they could survive and reproduce in it successfully (and let us not underestimate the incredible degree of complexity that is required in order for a species to be able to reproduce successfully, and to survive reproductively over time).

Why is it a matter of chance or luck that the mice ended up in just the right environment? Coyne believes it is a matter of chance because the causal sequences that produced the environment could have been different; there might have been no soil erosion that produced the sand dunes, no wind that moved seeds that produced plant food and shelter, no favorable climate that allowed mice to keep warm, and so forth. It is clear that he has misread the situation here, and he is not alone. In the first place, I have argued that the mutation could not have been otherwise because its cause—the cause of this actual mutation—could not have been otherwise, and the cause of this cause could not have been otherwise, and so on back to the beginning of the causal chain. This is also true for the interaction of the various causal chains that came together to lead to the mutation as an effect. Given the chains of prior causes, the subsequent event has to happen. This explanation also applies to the environment in which the organism lives. It too had to come about in just the same way, given prior causal conditions, and causal chains. And we can also say that two (or however many) causal chains *must* interact with each other, and produce a certain result, given the initial conditions, just as glaciers interact with the ground terrain to produce a river—the glacier had to happen given its prior causal chain, as did the ground terrain, and one must interact with the other, and produce a certain effect, viz. the river; there are no coincidental causal interactions in nature. Each causal chain has a separate causal explanation for its origins (though the origin of organic living things is much more complicated and interesting than the changes that occur in the realm of physics regarding physical matter and energy, and the fact that various living organisms come about this way, and interact with the environment in the way they do, is on a different scale—a difference in kind, perhaps—from the interaction of two non-organic causal chains with each other). But the overall point to take special note of is that it is fascinating to realize that nature has produced, by means of a deterministic process, organic, living things that can survive in their particular environments. The argument then that it is all just a matter of chance is not true, and even if I am wrong about the deterministic nature of the universe, the argument that the species just so happened to come about by chance is still extremely implausible (I will return to this form of the design argument in Chapter 8).

Coyne and others sometimes appear to accept at least a part of this argument because they are prepared to acknowledge that, strictly speaking, it is not true that in evolution the process is totally governed by chance. This is because they recognize that an organism is born into, or later develops various mutations in, an already existing specific environment, and its development is therefore constrained by this environment. Dawkins has expressed this point by noting that natural selection is the "non-random survival of random variants."[4] This means that the environment into which an organism is born places a limiting factor on the manner in which the organism survives, flourishes, and reproduces. This is true enough as a local explanation but as an ultimate or comprehensive explanation for how an organism came to exist in the first place, it is not true, because Dawkins and other evolutionary biologists conveniently forget that, on *their* interpretation of chance in nature, *the environment itself also comes about by chance* (because the prior causes of its effects could have been otherwise, just as they could have been with mutations). This means that the environment could have ended up with any nature whatsoever, perhaps only constrained by the type of matter present in the universe, and by the laws of science. On this interpretation of evolution operating with a large element of chance, no organism had to be born into the particular environment it was born into; it could just as easily have been born into a different environment (since this environment comes about by chance), and so this would affect its survival chances. Therefore, Dawkins is wrong to describe natural selection as a process involving a "non-random" survival, especially when referring to the comprehensive (or ultimate) explanation of how organisms and species came to be because this sounds as if he thinks the environment is *necessary* in some way, yet it is not on his view. It came about by chance, just as everything else did.

Ernst Mayr's work shows a similar confusion. Mayr believes that natural selection is the product of both chance and determinism. He holds that when we are considering all of the causal processes that lead to the production of a new zygote (including meiosis, gamete formation, and fertilization), "chance rules supreme at this step, except that the nature of the changes at a given gene locus is strongly constrained."[5] But when the organism begins to interact with its environment, this second step "is a mixture of chance and determination" because the environment limits an organism's adaptability and survival prospects. Mayr then adds an important qualification, which gives us an insight into his thinking on chance and causation; he notes that even at the stage where the organism comes into interaction with its environment, the outcomes are somewhat probabilistic because "natural catastrophes … may kill otherwise highly fit [i.e., adaptive] individuals."[6] He gives the example of the human eye and argues that it is mistaken to say that it originates completely by chance. He argues that it is the "result of the favored survival of those individuals, generation after generation, who had the most efficient structures for vision." He adds further that selection is also *not* teleological because there is "no known genetic mechanism that could produce goal-directed evolutionary processes."[7]

Applying my previous arguments, Mayr is incorrect on all of these points. He is wrong that the mutations occur by chance, in the way I have already explained; he thinks the causes of mutations could have been different but simply overlooks the fact that they could not be (*given* their actual prior causes, and the initial starting conditions). He is right that the environment constrains what natural selection can do with an organism, but wrong to describe this as a kind of determinism (*on his view*), because, on his understanding of chance, the environment came about purely by chance as well. This means that the environment could have had a completely different structure than the one it actually has. It is true that, given its *present* structure, it constrains the process of natural selection, as the organism interacts with its environment. But it is also true that the environment could have had almost any other structure than the one it actually has, if it came about by chance, and so the type of constraint that it places on the development of the eye, for example, is also purely a matter of chance. On my view, *both* processes are determined—the mutation process and the nature of the environment—and therefore Mayr's conclusion that nature is not goal-directed becomes controversial. *The causal chains themselves, operating from the initial ingredients, are moving nature toward teleological goals.*

There are two other ways to make this point. One is that it is simply incredible, as many have argued, that two chance processes come together simultaneously to produce the remarkably complex organisms that we find in nature. (This is why many think that evolution shows evidence of design; not because, as Paley is often read as saying, that structures like the eye, the heart, skeletal features, and the structure of habitats seem designed, but because the emergence through the process of evolution of complex species, culminating in the arrival of conscious observers like us, with consciousness, rationality, and moral agency, looks designed. It is this type of design that is suggested by evolution itself; it is this design that makes us doubt that it came about by chance—because the evidence from evolution itself is against it.) We can put the point this way: how can one tell if a process in nature is teleological, is moving toward a goal? One obvious answer is: if it looks like it was designed. One cannot reply to this objection about the emergence of complex species by saying that the process is not designed but operates by chance, because the process involves no elements of chance (so I have been arguing).

A second way to make our point here, and now arguing from the point of view that I am right in my claims about chance and randomness, is to emphasize that it is very interesting that a deterministic process gave rise eventually to conscious observers, given that there is no chance involved anywhere in the process. This prompts us logically to ask how the process got started in the first place, so that it leads to this outcome, just as we ask how the pool table and balls in our previous example got set up in a particular way. This raises the question of a possible designer. It would be quite remarkable, in short, if a process operating largely by chance led to the emergence of complex forms of conscious observers, but it is not quite so remarkable if the process is not governed by chance but is completely determined in the way I have been arguing for, so that, *given* a particular starting point, a particular ending *has* to occur. But the interesting question this prompts us to ask is: what is the cause of this initial starting point, the cause that leads to such amazing effects? Mayr is also importantly wrong in his claim that natural selection works somewhat "probabilistically"; the probability is made possible he thinks because something like a hurricane *could* come along and wipe out a food source, which would then affect the evolution of a particular species. It is incorrect to describe a happening like this (this kind of cause) as a "probable event," because this again implies that it might not have happened. But on a deterministic view of the universe—which is what we are working with at the practical level of modern physics—it had to happen this way. To say it is probable is to confuse our knowledge of reality (specifically of causes, causal chains, their interactions, and their effects) with reality in itself, to confuse our inability to predict what will happen with what actually does happen. This is a very common confusion in the discussion, and one I will return to later in this chapter.

Replaying the tape of life

We can now return to Gould's interesting and very influential metaphor concerning the tape of life, and reread it in the light of our arguments. If we examine Gould's point about replaying the tape of life in the light of our arguments about chance and determinism, we have to ask his question more precisely. We are not now simply asking if we replayed the tape, would we end up with the same organisms and species (and the same everything, including in the realm of geology and the environment, broadened to include galaxies and planets, and the whole universe) that we now have? Instead, we must ask: if we replayed the tape, *with the same starting conditions, ingredients, laws of science, timeline, etc*., would we end up with the same organisms, same species (and the same everything), that we now have (leaving human free will, or any direct intervention from an outside agent into the causal chains, out of consideration for the moment)? In other words, suppose the tape of the universe were to be replayed starting from exactly the same initial state—exactly the same big bang—would we end up with the universe having the same history up to at least the appearance of *Homo sapiens*? I think the answer to this question is yes. To invoke our example of the pool table again, it is like asking if we set up the same pool table with exactly the same type of balls all in the same places, and then hit them with the cue ball at exactly the same angle, same speed—everything the same, in short (including the environment of the room)—would they all end up in the same locations on the table as before? Again, the answer is yes. It seems to me that this is exactly how physics works, and why we can have confidence that happenings in nature (and indeed in the area of technology) will behave today the way they did yesterday, will behave the way they are supposed to behave most of the time. This is true unless the variables in any causal process under consideration change, which only means that the causal conditions change. But no prior cause can change, or be otherwise than it was (B must hit A at a precise angle and speed), given *its* prior cause, which also cannot change. Each effect must be what it is given its prior cause, which suggests a determinism about nature.

This is why events in nature can be explained in terms of the laws of physics, and their causal conditions; this is also why we can work out the cause if we know the effect, and why we can make predictions to what the effects of causal actions will be. This is, for instance, why we think we can predict what will happen to the environment if a certain type of global warming were to take place in a certain way over a specific time period; this is why we can say what will happen when we ignite a rocket engine on the moon, for example. We could not say what would happen with regard to global warming if some of the future causes in the causal chain (of which global warming is the end result) are subject to *chance*. If there is a true chance element pervasive throughout causation, it would be very difficult, perhaps impossible, to make *reliable* predictions.

Let us keep in mind that we are not simply saying that one of the causes/effects in any future causal chain that we make a prediction about might in fact not occur. This would only mean that we could not quite *predict* the chain with complete accuracy because there are too many variables involved, too many contributing causes to specify precisely. And if one of these causes occurs differently than what was predicted, then the end result will be different, of course. But these causes must still be what they are given their prior causes, in the way we have explained. So the overall conclusion is still the same. Gould, however, is clearly implying in his understanding of the process of evolution that a particular cause in the causal chain could have gone differently than it did in fact go. This is what he means when he says that even though a small change in the causal process might appear trivial at the time (say on a specific day a strong wind blew down plant cover and so larvae around the plant did not survive, or had their DNA altered), it could affect very significantly the long-term evolution of a species, the lineage of various species, and indeed the whole path of evolution, more generally. If we rerun the tape again, according to Gould, "Little quirks at the outset, *occurring for no particular reason*, unleash cascades of consequences that make a particular feature seem inevitable in retrospect. But the slightest early nudge contacts a different groove, and history veers into another plausible channel, diverging continually from its original pathway."[8] But he is overlooking the fact that no cause in the chain *can* be different, given its prior cause(s), that there is always *a particular reason* for an effect. There is always a causal reason for why a plant grows up on one side of a rock rather than the other, for why an apple falls from a tree and lands on a stone, for why an insect lands on a particular spot on a wall, for why a rabbit runs in a certain direction in a forest on a certain day, for why a fox eats a chicken; none of these events is "random" in the sense in which the term is frequently used in evolutionary biology. This is a conclusion Gould likely accepts in physics, but seems to forget about when it comes to biology (at very least I doubt he wishes to commit to the position that there is a significant element of chance operating in physical laws, and he does say that he thinks the origin of life was *not* a matter of historical contingency, but a "chemical necessity"[9]).

Evolutionary biology operates with an assumption of chance, whereas physics operates, especially at the practical level of everyday physics, with an assumption of determinism. I think that it would be quite possible for biologists to change their approach, and operate with an assumption of determinism.[10] This assumption would not fundamentally alter our understanding of how evolution operates, but it obviously would have significance for the philosophical and theological implications we draw from the theory. The assumption that chance and random elements operate in evolution, and the adoption of this assumption in the teaching of the theory, seems to me to be a current prejudice of the theory's main exponents, and not an assumption based on scientific evidence. Indeed, the scientific evidence from physics is significantly against a conclusion of chance operating anywhere in nature and in the universe. But if biologists were to work with a version of determinism about the operations of nature, this, of course, would open the door wide to the notion of teleology because a deterministic view of nature suggests that nature has teleological goals. It is also important to point out that not all evolutionary biologists dismiss determinism; one who moves in this direction with his fascinating theory of convergence is Simon Conway Morris, who argues that the fact that similar structures (e.g., the phenomenon of flight, and the structure of the eye) have evolved in nature *independently* several times is evidence that the tree of life may be more deterministic than is generally thought. Although we do not have space to discuss this argument here, Morris's conclusion is that you can "rerun the tape of life as often as you like, and the end result will be much the same."[11] We should also note that, despite what we may have been led to believe by some leading thinkers, there is no consensus about the issue of determinism within biology, at least as it might apply to the question of whether mutations are teleological in character.[12]

It is worth taking a moment here to consider how Gould might reply to our criticism as a way of helping us to see as clearly as possible the points at issue. Gould might respond, with regard to the beaver example in the previous chapter (where we noted that the chemical in the water from which the beaver drank caused a mutation in his DNA), that the chemical might just as easily not have been in the water when the beaver drank from it. Why is this? He might say because the plant that secretes the chemical might not have grown on the river bank. Why not? Because the seeds that led to this plant were scattered by the wind from a tree in the middle of the forest and it was only by a lucky coincidence that they landed on the river bank. They could just as easily have landed anywhere in the forest. But some of them just happened to land near the river bank, take root, grow well, and eventually secrete a chemical into the water. However, this kind of explanation as a reply is incorrect because I am arguing that the seeds *had* to land where they did given the prior cause of their movement, which would involve a certain wind speed, certain wind direction, unobstructed path, etc. *Given* all of these causal factors, together with the physical laws of the universe, the seeds *must* land where they do. Suppose Gould replies that the wind speed could have been different on the day the tree released the seeds. Again, this is not true, for the wind movement has a prior cause, and this cause will produce wind of a certain velocity at a particular time. Only if this prior cause were changed, could the end result be changed, but this prior cause cannot be changed, *given* its prior cause. Suppose Gould were to suggest that the trees that provided the shade necessary for the plant to grow might not have grown in just that spot. Again, this objection is off the mark, because given their prior causes, they must grow in that spot! (This analysis also assumes that the beaver's behavior is completely determined by his environment and internal structure {by physiological and neurological structures in the brain}, but I will not apply this deterministic argument about animal behavior and action to human behavior, as we will see later.) This is how it works in physics, and so it must be also in biology.

The same line of reasoning applies to the causal chains present in evolution itself. And before we illustrate this point with an example, it is worth recalling that at least some evolutionary biologists believe that biology is reducible to physics and chemistry, a conclusion that would strengthen my argument (though I believe they are wrong about this claim!). To say that biology is reducible to physics and chemistry means that every event in biology is fully explicable in terms of matter and energy and the laws of physics, that biological systems are simply a subset of sophisticated types of physical systems that are still subject to the same rules of cause and effect as all physical systems are (such as volcanoes, the formation of glaciers, the earth's atmosphere, and the development of a habitat). So if one held this view, it would be odd to claim that there is a large element of chance involved at the biological level that is not present at the level of physics and chemistry, for I do not think there are many scientists who believe that chance occurs in physical processes *at least at the working level of science*. Every scientist operates with a deterministic view of the physical world at the working level (irrespective of which theoretical perspective they might favor); indeed it is hard to see how one could do science if one did not operate with this assumption. Otherwise one would have to say that in any causal chain under discussion—say one dealing with the effects of global warming—that the results of one's research could all be undermined because an effect could occur somewhere in the chain *by chance*. And let us not forget, as we just noted, that this does not mean that some other (hidden, unknown) variable could come along that we did not know about and that could influence the end result, for this is only a point about our *knowledge* of the causal chains, and about *predictability*, as I have pointed out, and is not an argument against determinism in nature. The deeper conceptual question as to whether there is any chance operating at all in physics, and specifically in the process of cause and effect, the question that philosophers are interested in when they are thinking about chance and necessity in reality, and the one that scientists might also be interested in when they are not doing science, can still of course be asked, and I will return to it again later in this chapter. But as I noted above about Gould's view, this deeper question is not one that evolutionary biologists are raising and taking a stand on when they are talking about chance and randomness in nature; in fact, the vast majority of them believe that the answer to the metaphysical question is that there is no chance in nature (and so no effect, or outcome, can be truly "random.") This is why many of them who are secularists and naturalists have such a problem with human free will—precisely because they have a deterministic view of how nature operates, and so they cannot see how free human actions could fit into a deterministic universe (Coyne is one thinker who contradicts himself on this issue).[13]

Let us return then to how Gould, or indeed any of the thinkers we have mentioned, might respond to my argument, with regard more specifically to causation in biology. Suppose an expectant mother smokes cigarettes and the smoke alters the DNA of her baby at a certain embryonic stage of development. Let us say that the smoke affects the development of the brain cells of the child, causing the child to have learning difficulties. Let us accept for the moment that the mother has free will and could have chosen not to smoke, and so we will leave this part of the causal chain out of consideration, and simply consider the causal chain from the moment the smoke enters her body. Gould would argue that it might have been the case that the membrane protecting the child was a bit stronger than it in fact is, and if so would not allow the chemicals in the smoke to penetrate, and so the smoke would have no effect on the development of the embryo. On my view this could not be the case because the membrane has a particular causal history that *must* result in the development of just this membrane with this degree of thickness, resistance, etc. Gould might argue that the chemicals could have penetrated the membrane but have been unable to penetrate the fluid surrounding the embryo, and so be unable to cause (detrimental) changes in the molecular structure of the embryo's cells, and so forth. I am arguing that none of these alternatives could happen given their causal history. It is only if the initial conditions are different that the outcome can be different.

All of this shows us that no process in nature, including evolutionary processes (the events that contribute to evolutionary change), happens by chance. There is no chance at all operating in the natural world, and hence there are no random events. Many thinkers have been sloppy in talking as if there is a major element of chance in the process. The only way to change the course of history is to change the initial conditions, which brings us inevitably back to the larger question of *how the initial ingredients came about in the first place, to the question of the origin and nature of the big bang*. And of course this brings us back to the question of an intelligence and a design behind the initial conditions, one that is perhaps responsible for the initial conditions. This line of reasoning also opens a clear way to develop an argument for the compatibility of religion and evolution.

Chance and probability

We now need to develop the above position in more detail by adding two very important qualifications regarding chance and probability, and regarding quantum mechanics. These additional points will help us to illustrate further our argument about chance and determinism. The first important qualification is that, just as we must distinguish carefully between chance and determinism, and need to be clear about what we mean by chance and randomness, we also need to distinguish clearly between chance and probability (or predictability). These two notions are *frequently* confused in this general discussion.[14] Perhaps a helpful way to think about this distinction is to ask the following question: what is the *probability* that a particular event will occur in nature, such as the formation of the Grand Canyon, a comet slamming into the earth, the occurrence of a particular mutation (such as that causing sickle cell anemia), a particular species coming into being, a storm happening that has a deleterious effect on the survival of an organism, or a particular environmental change that affects species adaptability, and so forth? On the view I have been arguing for, the probability of any particular event occurring is 100 percent, *given the starting conditions, and the laws of science*. There is no possibility of the end result—the events, or the causal effects—not occurring, given *their* prior cause or causes, and there is no possibility of *these* causes not occurring, given *their* prior causes, all of the way back, as we have explained, to the beginning of the causal chain(s). Therefore, the physical universe is completely determined (leaving aside free will and the possibility of miraculous intervention). However, it does not follow from this determinism in nature itself that we humans can *predict* which events are going to occur in the future from our studies of sets of present causal conditions. This is because there are too many variables involved in each causal chain; it is not only too difficult to discover the actual number of variables involved in many causal chains, it is also too difficult to measure the behavior and effects of these variables with precision and certainty in most cases. As noted above, we continually confuse *our inability to describe how something happened, or to predict what will happen*, with the view that when it happened it happened by chance (i.e., that its prior cause did not occur by necessity). This is the key point.

The confusion is evident in many writings on evolution. For instance, Coyne introduces the example of tossing a coin, and notes that there is a 50 percent chance of getting heads on any given toss, but if a person were to make four tosses of the coin, there is only a 12 percent chance of getting all heads or tails. He is using this example of the causal behavior of the coin as an analogy for how probability works in nature, specifically in evolutionary biology, saying that "especially in small populations, the proportion of different alleles can change over time *entirely by chance*."[15] As noted above, many discussions of the process of mutations in biology, and other causal influences on development, are often illustrated by a calculation of the probabilities involved in an organism or species getting a particular mutation or being influenced by other causal factors, which subtly emphasizes the supposed *chance* nature of the process, and so the *random* nature of the outcomes.

However, there are significant problems with Coyne's example, problems that apply to all examples of this type. The first is that human free will plays a role in the tossing of the coin so that we are not working with a completely closed causal chain; so the end result is distorted, and not indicative of how events would behave in nature where there is no free will (of course, if human beings do not have free will, then we would need to analyze coin tossing in exactly the same way as any other causal process in nature). But, second, and more important, the probability of whether we get heads or tails is only a reflection of our ability to *predict* which coin face will come up, it has nothing to do with which face actually comes up. To say that an event is "probable" is to report on the likelihood (*from our vantage point as human observers*) of it happening; the reason we can only report on its likelihood is because we cannot be precise about the full causal conditions and causal chains that produce it; but if we could be precise about these, we could then predict with certainty which event would occur. In the case of coin tossing, a calculation of probability means that if we predict that heads will come up, there is a 50 percent chance *the prediction* will be correct, but it does not mean that what comes up (in nature, outside of our prediction) has a 50 percent chance of occurring! (The prediction represents an *epistemic* uncertainty or indeterminacy, on the side of the knower, not an *ontological* uncertainty or indeterminacy, on the side of reality, outside of the knower.)[16] Which face comes up in a coin toss is 100 percent determined by cause and effect (leaving free will out of the causal process for now)—only our ability to predict the outcome is what is measured when we talk about probability. Probability only makes sense in relation to human beings; there really is no such thing as probability in nature—there are only initial chains of causes leading to necessary end results. If there were no human knowers, there would be no probabilities. Indeed, talk of probability (probable happenings in nature) only makes sense if we assume that we can predict reasonably accurately what way nature will behave, given our knowledge of causes and the laws of physics, and this requires that nature be deterministic.

It is true that sometimes we confuse the two when we speak (too loosely) about probability in ordinary life; we sometimes seem to be saying that the probability describes what happens in nature rather than describing our attempts to predict what will happen based on our incomplete knowledge of the causal chains. However, probability itself plays no *causal* role in nature. A little refection shows that we must maintain the distinction between our prediction about what will happen and the way in which causally it actually happens in the universe. This is why, for instance, in coin tossing we often try to achieve exactly the same height with our next throw as we reached with our previous throw—because we think we may get the same result! Why is this? Because we think that we may repeat the *exact causal chain* in the second case that we had in the first case, and if we do, *we believe we will get the same result*. As noted above, we know that this is how physics works. It is how physics is understood and practiced; it means that nature is determined, and it also must apply in all areas of nature, including biology. So we must be careful how we use concepts like "random variation," "random sampling," "random walk," "genetic drift," etc., and understand them as describing the imprecise nature of our calculations rather than the way nature behaves in itself. This is why Laplace noted that an infinite mind that knew every prior ingredient and cause in the universe should be able to predict the future of the universe with certainty, to be able almost to "see" it present all at once.[17]

To his credit, Monod, unlike Coyne, does see this distinction, but fails to draw the right conclusion from it. Monod recognizes that, speaking strictly, it is necessary to remove from consideration of these games of chance, such as playing dice, or cards, or slot machines, the human factor. So he suggests that we could imagine a *machine* (thereby leaving human beings and specifically the question of free will to one side) that tosses the coin in exactly the same way each time, but he does not think this kind of mechanical precision holds in nature, and suggests that "between the occurrences that can provoke or permit an error in the *replication* of the genetic message and its functional consequences there is also a complete independence [meaning the process is not aiming at a particular goal]," and goes on also to invoke "uncertainty [indeterminacy?], embedded in the quantum structure of matter."[18] (I will consider quantum theory below, as a separate issue.) But Monod fails to see that the end result is nevertheless determined, and cannot occur by chance. Indeed, this is how the term "loaded dice" is to be understood—it means that we have designed them to produce a different result than the usual result; we have added a mechanism that allows us to *control* the causal chain rather than nature controlling it; but the result from loaded dice and the result from unloaded dice are nevertheless both determined, and are both a consequence of specific causes and their effects! Both results are determined; it is just that in one (loaded) case we know the causes better than in the other (natural) case, and so we can "predict" (i.e., *describe more accurately*) what the end result will be. If there was true chance in nature, none of this would be possible; and "loaded" dice might still not produce the desired outcome because a chance occurrence that we could not control for in the causal process could occur at any moment. (Of course, loaded dice do not in any case produce the desired outcome every time. Why is this? The answer is yet another example of deterministic causation at work in nature.) This is also why we seek the causes of mutations in nature; so that we can control them, either by direct intervention or by drugs. Both procedures, and indeed the whole of medicine, assume a deterministic universe.

Coyne applies the coin analogy to the causal activity involving alleles (different forms of a gene that can lead to different observable traits in an organism), and says the following:

> Every individual has two copies of each gene, which can be identical or different. Every time sexual reproduction occurs, one member of each pair of genes from a parent makes it into the offspring, along with one from the other parent. *It's a toss up* which one of each parent's pair gets to the next generation. If you have AB blood type, for example (one "A" allele and one "B" allele), and produce only one child, there's only a 50 percent chance it will get your A allele and a 50 percent chance it gets the B allele... The upshot is that, every generation, the genes of parents take part in a lottery whose prize is representation in the next generation... This "sampling" of genes is precisely like tossing a coin... The proportion of different alleles can change over time *entirely by chance.*[19]

I think we can clearly see the confusion in this view of evolution and chance. Coyne is right to say that it is "a toss up," if he just means that we can't predict in advance (because of our lack of knowledge) which parent's allele the child will get. But he then mistakenly thinks this means that *the allele the child actually gets is due to chance.* He will acknowledge that to say that the allele is passed on by chance does not mean that there is no cause for which one is inherited. He realizes that there must be a cause for why one particular allele is passed on and one is not (not just one cause, of course, but usually many; a set of causal chains is involved). Coyne, however, thinks that whatever the cause is, it could have been different, and this is where he makes a mistake. I have tried to show that the cause *could not have been otherwise,* could not have been different given its prior cause, and so on back to the beginning of the causal chain(s). There is no lottery in nature itself, where everything is determined by the initial conditions and the laws of physics, but it might seem like a lottery to us since it is very hard to predict how future causal chains will go. It is easy also for us to confuse our inability to predict what the causes are (of the various causal chains and their interactions, perhaps impossible to predict in practice, but not in principle, of course), with the claim that there are chance occurrences in nature itself, a mistake that seems to me to be widespread in evolutionary biology.

Monod makes a similar mistake in his discussion of chance operating in nature, and confuses chance with unpredictability in his examples of bacterial reproduction and in an organism's system of defense through antibodies.[20] We can illustrate further with another example of breeding fish. Monod's view, and the view of most evolutionary thinkers, is that if you have four different fish that interbreed together, that there is a large element of chance involved in which offspring get which characteristics.

This is because all sorts of things could cause or prevent one fish from mating with another, and it also means that in the process of reproduction many different outcomes are possible because the causal chains operate by chance, in Gould's sense. I believe this understanding of how nature works is mistaken. The only uncertainty in the process is on our human side; there is no uncertainty on the side of nature. It is only that we can't predict which offspring will get which characteristics because we don't have sufficient knowledge of the enormously complex causal chains involved (despite us having a good outline of their general structure), but the whole process is determined from start to finish in terms of the biological makeup of the fish, their environment, and the interaction of the two. To believe otherwise is simply to reject modern science. And "artificial selection," selection by design to produce desirable characteristics in animal or plant species, is simply the application of human free will and human knowledge to otherwise deterministic causal chains. This means that if a fish breeder took three or four fish and had a very good knowledge of the causal processes involved in fish reproduction, she would be able to bring about a desired result in the offspring—this is precisely because the process is deterministic, and does not involve elements of chance.

Biologists will sometimes reply to this argument by contending that one can refute this conclusion empirically by carrying out an experiment involving the breeding of fish, and then running the experiment a second time in identical conditions. But the second time around one gets different results, and this means that an element of chance is involved, and that the process is not deterministic. However, this objection does not work for the simple reason that the experiment is *not* repeated in identical conditions because it is too difficult in a biological experiment of this sort to reproduce exactly the same conditions. There are always some variables that one does not have complete knowledge of, and control over, and there are always unknown, hidden variables that one cannot control for, but that influence the final outcomes. But if we could reproduce exactly the same conditions the second time around, who can doubt that we would get exactly the same results? This is because identical conditions and identical causes produce identical effects. Isn't this true in the area of physics where identical conditions are easier to achieve, for instance in the utilization of electrical circuits, and isn't it true in the area of medicine? Isn't this what we are aiming at when we try to develop a cure for a disease? The answer is yes in all of these cases, and the answer is also yes in biology. Darwin may have also appreciated this point because he once observed that "chance" was just another way of saying that we don't have complete knowledge of the causes of things and of the laws of physics.[21]

Chance and quantum mechanics

Let us now turn to one interesting objection that is often proposed to the view I have been defending that causal forces in the universe are deterministic. This is an objection that is based in the twentieth-century theory of quantum mechanics. Quantum mechanics deals with the nature and interactions of subatomic and atomic particles, such as the quanta, neutrons, electrons, quarks, and photons. Particle physicists study the energy and momentum of these particles, among other things. Some of their findings are unusual and challenge us to think about the nature of the microscopic world, as well as about the notions of chance and necessity. Although many aspects of quantum theory are still speculative and not fully understood, there is no denying that it has been a very successful area of research, becoming the basis for transistors, lasers, modern electronics, and nuclear energy.[22]

Researchers note, however, that one runs into difficulties when studying subatomic particles, difficulties that do not arise in the ordinary macroscopic world of our experience. One main difficulty is that when we study a subatomic particle, because it is so small, it is affected by the *methods* we use to study it, specifically by the energy from the light source (light energy does not affect our study of large objects, such as tables and chairs). This means that when we are studying the particle as observers, we are also changing it, affecting how it behaves. The energy from the light wave that is used to study an electron, for example, will affect the *motion* of the electron in a way that cannot be predicted; if a longer light wave is used to avoid this problem, then the measurement of the *position* of the electron will be affected. In short, it seems that we cannot know both the position and the rate of motion of the electron at the same time, a conclusion called "the uncertainty principle" by German physicist Werner Heisenberg. A consequence of this state of uncertainty is that we can't state the laws concerning the behavior of particles operating at the subatomic level with precision; they have to be expressed as statistical probabilities. So in this way they differ from the laws governing macroscopic objects, which, although arrived at by induction, are nevertheless accepted as laws in the sense of holding in every given instance.

What are some of the implications of these findings in quantum mechanics for the conclusions I have been drawing in this and the previous chapter? First, some may appeal to quantum mechanics to argue that the uncertainty principle adds an indeterminacy to nature itself, and perhaps this indeterminism could be appealed to in some way to undermine my earlier argument that nature is deterministic. The question is whether the indeterminacy we observe, and that frustrates us in our attempts to state the laws that govern subatomic particles in precise terms, exists *in the particles themselves*, or is it only a result of the difficulties we find in *our attempts to measure* these particles at close quarters? The theory of quantum mechanics gives rise to a fundamental philosophical question concerning how we are to decide if what we are observing in the electron, for instance, reflects its actual behavior in itself—the way it really is in objective reality—or whether it only reflects how it behaves when *we* are observing and studying it? There are several schools of thought on this matter within quantum mechanics, several different interpretations of the uncertainty principle. One is that the indeterminacy is in the particles themselves, and is therefore an objective truth about reality, which was Heisenberg's view (this view is called the Copenhagen interpretation). A second view is that the uncertainty results only from our inadequate current attempts at measurement, the view of Einstein and Max Planck, and more recently of physicists like David Bohm. As Einstein famously noted in his rejection of indeterminism in nature itself, "I am convinced that [God] is not playing at dice [in the universe]."[23] My view is that the thesis that nature is deterministic is the most rational one to adopt on the grounds that all of our experience is against the view that the behaviors of particles could be indeterminate in themselves. From a logical point of view, it is not clear that such a notion is even intelligible. And if a notion seems to be logically unintelligible, we should be very reluctant to accept it.

A second important point is that it is still the case that the various probabilities concerning the laws governing subatomic particles at the quantum level cannot be used to rule out determinism at the quantum level. What this means, as we noted above in our discussion of probability, is that even though our *statement* of laws only captures probabilities with regard to the behavior of particles, we cannot conclude from this that the actual behavior of the particles is indeterminate; all we can conclude is that our *knowledge* of the causal factors is incomplete. Of course, it might nevertheless remain the case that the world of subatomic particles is indeterminate from a practical point of view in the sense that we will never be able to overcome our problems in trying to measure and study these particles, that our techniques will always affect our results, and this would mean that our knowledge of this microscopic world would remain incomplete. But this would be due to a barrier facing our attempts to discover knowledge of the subatomic world, not because this world in itself acts in an indeterminate way (i.e., does not follow the laws of science *all the time*).

This leads to our third point that at the atomic and larger levels the world *does* operate in a deterministic way, and this allows us to pursue objective science in a realist sense, to discover and apply laws that hold consistently, and to arrive at certain knowledge. It is the case that any indeterminacy at the subatomic level, which is hotly disputed, does not transfer over at the atomic level (at least most of the time) because the effects are so small. As Timothy Shanahan has expressed it, "in all contexts outside of the peculiar domain of quantum mechanics we are justified in assuming that identical causes result in identical effects."[24] It is also the case, very importantly, that the uncertainty does not even affect most of the conclusions of quantum mechanics itself, and this is why we can use our knowledge of this realm of reality to build technology—for example, lasers—because these technological devices consistently follow the laws of physics. The indeterminacy that is supposed to be possible at the quantum level does not significantly affect our ability to predict the behavior of particles in consistent ways, which one can take as a further argument that the behavior of particles is in fact operating according to the determinism I have been defending, that contained in the relationship of causes, and causal chains, to their effects.

The behavior of quantum particles might provide a way to argue that there may be some kind of genuine chance present in the universe at the subatomic level. This is because, supposing for the sake of argument that the particles themselves are indeterminate in their movement and/or position (and perhaps in their nature too, if their nature is significantly affected by their movement and position), this could perhaps carry over in some way in some instances to the macroscopic level. Perhaps then this behavior could affect the behavior of larger objects in such a way that some of their behavior could be described as "subject to chance," that is, these objects would not behave in a deterministic way. However, it is hard to grasp what this means. Perhaps it means that when we light a match and hold it to paper, the quantum effects going on at some level in the flame might cause it to not light the paper, even though all other things that need to be present in the causal chains are present (all other things are equal). And so this would mean that, *although all other things are equal*, that on some occasions the match would not light the paper (or your car engine would not start, etc.). It seems to me to be very difficult to accept that chance could come into the universe in this way. Moreover, it would be hard to detect it in the sense that we could not distinguish between cases where the scientific law does not hold because of quantum uncertainty/indeterminism, and cases where the effect did not occur because all things are, in fact, *not equal* in a particular instance (for example, the breeze might be too strong, there is not enough oxygen, the paper is compromised in some way; in short, there are hidden variables involved at the macro level of ordinary causation, a problem that suggests that it may be impossible *in principle* to establish indeterminacy at the quantum level). However, this is one interesting way to speculate about how perhaps a possibility of genuine chance (and not just an appearance of chance because of inadequate measurement, or imprecise probability calculations, all of which come from the human side) could enter into the causal history of the universe. This also might be a way to further develop the metaphysical question of whether everything in the universe has to happen the way it does happen.

Nevertheless, this line of argument will not help those who want to argue that this is the way chance and randomness work in the process of evolution. This is because when evolutionary biologists like Gould, Dawkins, Coyne, and others we have discussed, talk about chance in nature, they are *not* talking about an indeterminism or an unpredictability (in principle) at the quantum level. This level does not usually enter into explanations concerning the process of evolution, which deals with macroscopic entities, like the molecular makeup of cells, the anatomy of species, fossils, habitats, environmental features, the causal interactions between all of these, and so forth. These evolutionary biologists are talking about chance and randomness in the way we suggested above, in the sense that *the causes of events* could have been different, that various events did not have to happen the way they did happen (both environmental events and mutation events), and so there is no outcome toward which nature is moving (no teleology). They hold that there is genuine chance or indeterminism in nature at the level evolution operates at, irrespective of the quantum world.

Therefore, the various species that emerged might have been different, as might their various structures along with that of their habitats (let us not lose sight of the fact that the habitats must come about by chance as well, according to their understanding of causation and chance). Of course, if determinism *is* true at the quantum level, as I have suggested above, then it would have to carry over into the world of ordinary physics as well, and so there would be no room for indeterminacy in nature. But even if there is some indeterminacy at the level of subatomic particles, there is still determinism at the atomic level, and this is the level evolution is operating at. In other words, the laws of physics still apply at the atomic level, and are extremely reliable, and so the process of evolution, like all processes, is still subject to them.

However, what these scientists are really talking about is *unpredictability*, not chance or indeterminism in nature. They sometimes appear to be suggesting that nature does not always quite follow the laws of physics, or to be suggesting that some events in evolutionary history do not have any cause, when in fact they should be more careful, especially given the serious implications of what they are saying. Gould, for instance, says that we cannot predict which species will emerge from the evolutionary process. This is true, but it does not follow from this that the species that did emerge did so randomly. We can't predict which species will emerge from the process for the same reason that we can't predict when a drive belt will break in our car: because the causal chain involved is too large and complex for us to gain complete knowledge of it, but we do not doubt that there is a causal chain, and that the breaking of the drive belt, or the emergence of species, is fully subject to the chain of prior causes. Gould's mistake is to think that the process that produced a species could have gone differently even if the initial starting conditions remained the same.

Is evolution progressive?

This brings us to a final interesting question to consider before we bring this chapter to a close, the question as to whether or not evolution is progressive. A number of thinkers have discussed this topic, including Darwin himself.[25] The question concerns whether evolution is moving in the general direction of producing more and more complex species, more sophisticated, higher forms of life? When we ask if evolution is moving in the direction of increasing sophistication, the question may be misunderstood. What we really mean to ask is whether the universe is moving this way, including our natural world; more specifically in terms of evolution, we are asking if the processes of mutation and natural selection are moving this way? But let us remind ourselves that these are not personal or impersonal *forces* of any kind. Sometimes we forget this fact, and talk as if natural selection is some kind of independent "force" operating throughout nature; getting past this misunderstanding will help us to appreciate more fully the argument in this chapter. Mutation refers to a process of cause and effect that takes place in DNA, just as erosion refers to a process of cause and effect that takes place with rocks and river beds. And natural selection describes how the environment changes over time, again due to prior causes, and how species interact with it, also due to prior causes. It is the process of cause and effect that explains how a mountain came to have the specific features it has, from the time the process began to any time later when we come to study it. The deterministic nature of change is responsible for the current structure of the mountain.

The progressive nature of evolution may be seen as the end result of the fact that the universe and nature operate in a deterministic fashion. Given the beginning ingredients, and the laws of the universe, the increasing complexity had to come about, including that found in biological life forms. And so there is nothing random about the species (and their natures) that emerge from the process. This progressivism presents a problem for those who say that evolution is significantly governed by chance, and indeed is another argument against this view. This is simply because if evolution is governed by chance, it would be very unlikely that it would be progressive. It would be more likely to be all over the place in terms of its outcomes, since not only could any event happen to retard the progress in the development of a particular species, and of species in general, both in the structure of the species and in the species' environments, but it is also much more likely that these destructive events *would* happen. Quite simply put, the progressive nature of evolution suggests a design in the overall direction of nature. Of course, one can argue that logically the progressive nature of evolution might be a coincidence, but this is not reasonable to believe. Is it believable that species are getting progressively more complex and sophisticated if each mutational and environmental change comes about by chance?

It is in response to this problem that some argue, as we noted above, that chance and determinism can both play a role in nature because any change at the mutational level is *constrained* by the environment in which it takes place. This means, as Coyne has attempted to illustrate, that only mice can emerge from an environment of mice, and not bats, so it would be expected that we might get more sophisticated mice (some of the mice might have a mutation that enabled them to survive a disease that was gradually wiping out all of the other mice, and so forth). The first part of Coyne's argument is correct (at the level of microevolution, but not, crucially, macroevolution) that one can only get mice out of mice, but his second point that it is to be expected that later mice would be more sophisticated than earlier ones is not convincing, once one assumes that the process is governed by chance. This seems to me to be a clear case of Coyne papering over problems in his explanation of evolution. If the process is governed by chance, it is to be expected that the mice would not survive, perhaps not even evolve at all, or that their lineage would regress in some way, not advance, particularly since the decisive mutations are all occurring *by chance*. So it does not follow from the fact that, if mice exist in a certain environment, we would be more likely to get progressively more sophisticated mice in a similar future environment, because, if the mice came about by chance and the environment came about by chance, then it is very *unlikely* that one would complement the other. This is also true for progress in macroevolution. If the process is significantly governed by chance, would it not be unlikely that we would see progressive complexity in the emergence of species, rather than occasional, or even frequent, regression (for example, a line of descent with apes, *Homo sapiens*, *A.africanus*, *H.erectus*, chimpanzees, in that order)?

One might reply that what we have here are only two aspects of the same process—that the fact that species can survive *means* that the environment must be suitable for them, but this argument is undermined by the claim that *any* one of a very large number of mutations could have occurred. So then we have to explain the likelihood of *continuously favorable* mutations occurring and conferring a selective advantage in an environment that is not itself preset but that originated from, and is always subject to, a process significantly governed by chance. One cannot reply to this line of reasoning that since the mice survived, the environment must have been suitable for it, or that each must complement the other. This is because we are talking about the *likelihood* of the mice surviving (and of progressive complexity emerging), given the presence of a large element of chance throughout the process. Whatever way one looks at it, the progressive nature of evolution is difficult to explain on such a view of chance operating in nature. Nor can one reply by saying that the present environment would be constrained by the previous environmental structure—this is because the previous structure *also came about by chance*; this argument also applies to present mutations. The fact that *each* successive step is significantly governed by chance, and yet the end result over time is continuous, progressive sophistication on both the side of the environment and on the side of life, with both plants and animals, is what is very difficult to explain on this view. This is why the progressive nature of evolution *suggests design, and therefore a designer*.

7

How Does God Act? The Compatibility of Religion and Evolution

It is time now to turn our attention more directly to the ways in which religion and evolution might be compatible with each other. We have seen in the preceding two chapters that the process of evolution does not contain any elements of chance. So we must draw out the implications of what this means for our understanding of the workings of evolution, its relationship with religion, and how the central issues all fit together concerning our understanding of the nature of human beings, particularly with regard to our special qualities of reason, free will, and morality. Our argument is that the universe behaves (outside of human free will and any intervention by God) in a deterministic manner. This means that from a given starting point consisting of a set of ingredients or initial conditions, together with the laws of physics, the events that subsequently occur due to causal interactions *have* to occur. This conclusion applies to *anything* that happens in the universe, not just evolution, but of course it applies to evolution as well. Given this, the question of how the initial conditions came about is obviously paramount. And, of course, as we will see, the initial conditions *had to come about somehow*!

So we need to do three things at least in developing this argument further. The first is to consider the ultimate question as to how the initial ingredients and the laws of physics, which express how the initial ingredients behave, came about in the first place. The second is to reconsider and revaluate those philosophical and theological implications identified in Chapter 4 that evolution appears to have for ultimate questions about the universe and about human life. What are the implications of my argument that evolution does not contain any elements of chance for the question as to whether the existence of the human species is necessary or not, or for whether the human species differs in kind or only in degree from the other species? And more generally, then, for the question of the compatibility of religion and evolution? Does the fact that the process of evolution involves no elements of chance (as does no natural process) mean that there is no chance at all operating at any level of reality? We should also ask more theological questions such as why would God create the species using a process like evolution? In addition, how are evolution and the Bible, in particular, compatible? Our third task is to look further at other issues that naturally arise out of this discussion, such as at least attempting to examine in more detail the ways in which evolution and religion might be compatible (rather than just noting that they are); and what the implications of my position are for our understanding of evil (particularly natural evil) in the world. Moreover, we should ask what role do distinctive human qualities such as higher consciousness, reason and logic, free will and moral agency play in our thinking about evolution, and in God's plans for humanity. We will address these fascinating but challenging questions over the next two chapters.

Ultimate origins: An argument against naturalism

Perhaps the most fascinating question of all is the question how did the initial ingredients, together with the physical laws, come about? Where did these come from? Sometimes we don't focus enough on the mind-stretching enormity of this question. It is important to emphasize that we are not seeking with this question what philosophers sometimes call the *local* cause of the universe, but the *ultimate* cause. The local cause could turn out to be that the big bang was caused by a prior physical event (the "little bang") that we previously did not know about; or, more speculatively, that our universe was caused by another universe, or by another cause from another dimension of reality. This is the kind of cause that science in its everyday practice would pursue. But the philosopher wants to know the *ultimate* cause, the deeper cause, of the universe; that is to say, how *anything got here in the first place*. We should distinguish carefully too between the question and the answer. Although the answer may be difficult to arrive at, the question is essential and one that anyone thinking about the questions of this book must ask, otherwise they are leaving out something fundamental. And this is where naturalism as an explanatory theory of reality exhibits quite a serious weakness because in its modern versions it fails to address this question adequately.[1]

The question I want to focus on in this section in particular is whether naturalism as an explanatory theory of reality is a reasonable position? Although there can be different ways to state the thesis of naturalism, and there are a few different versions of it, the basic claim behind all naturalistic alternatives to religious belief is that everything that exists is either physical in nature or depends upon the physical for its existence, and so everything consists of, or came about from, some configuration of matter and energy. This view is reductionistic in that its proponents hold that any feature of the universe that appears to be non-physical, such as human consciousness, can still be "reduced" to (i.e., can be explained in terms of) matter and energy. Naturalists also usually hold that all prior causes for any event that occurs in the universe are physical in nature. So naturalism implies then that there is nothing supernatural that exists, nothing non-physical exists in its own right as a basic element of reality. It does not necessarily imply, though this is how it is usually understood by most of its proponents, that there is no designer of the universe, but it would mean that if there is a designer of the universe, or a designer of life, the designer is physical in nature, and so, like all things of a physical nature, would have to have a physical cause. I wish to suggest in this section that this naturalistic explanation of reality is not reasonable to believe.

It is not my intention to defend this claim by offering arguments for the existence of God. I believe such arguments are strong, and the next step of any argument against naturalism, including the one I offer here, would be to develop them. And sometimes philosophers do argue that the reason that naturalism is irrational, or not believable, or not a good explanatory theory of reality, is because they have a better theistic theory. This approach is a very strong one, and, along with many philosophers, and other intellectuals and religious believers, I endorse it.[2] But I shall not pursue it here. All I want to emphasize is that, leaving any arguments for the existence of God or for the reasonability of the religious worldview completely aside for the moment, and taking naturalism purely on its own terms (and not in relation to the theistic alternatives), it is a view that no reasonable person should accept. To see why, let me lay out the main structure of a typical argument for naturalism. The argument is usually based on inductive reasoning, and goes like this:

1. Most events in the universe that we observe and study are physical and come about by means of physical causes.
2. Therefore, (it is very likely that) all events are physical and have a physical cause.
3. Therefore, (it is very likely that) naturalism is true.

But if one accepts this argument, then one must also accept the following claim:

A. The origin of the universe is physical in nature (from 2 and 3 above)

But the problem with accepting this claim is that one cannot explain how the origin or cause of the universe came about on a naturalistic view. This is because:

4. (It is very likely that) Every physical event has a physical cause (from 2 above)
5. The origin of the universe is a physical event
6. Therefore, the origin of the universe has a physical cause

But from this line of reasoning, a problem emerges for naturalism:

7. The origin of the *cause* of the universe (and of any prior causes) must have a physical cause
8. A chain of physical causes, whether finite or infinite, cannot provide an *ultimate* explanation
9. Therefore, the *ultimate* cause or explanation of the universe cannot be physical

But:

10. The universe must have an ultimate cause or explanation
11. Therefore, naturalism is false.

The first argument is based on inductive reasoning from the study of causes and effects in our universe, and then generalizing to a conclusion of naturalism. But the argument carried to its logical conclusion presents problems, which emerge in the third argument. Premise 8 leaves us with what philosophers call an infinite regress, meaning that one has to keep appealing to prior physical causes to explain the existence of the present physical effect, but this appeal goes on indefinitely backwards into the past with the result that one *cannot* explain ultimately the present physical effect. This means that if one holds that everything that exists is physical in nature, one will be unable to explain why there exists a universe in the first place. And the crucial point is that one *must* explain this fact (premise 10). And so it follows from this that in order to explain the origin of the universe, the cause cannot be physical in nature (proposition 9), because if it is physical in nature, it too would need an explanation. So one has to go outside the physical order to explain the universe, and so naturalism is false (proposition 11).

Faced with the problems presented for their position by these arguments, naturalists will often try to question premise 4 (and/or premise 10), at least as applied to the first event in the universe. William Rowe has suggested that our desire to find an ultimate cause is based on the universal human desire to find a reason for everything in the universe. This is motivated by the natural curiosity of human beings who want to know how things work, but he argues that it does not follow from the fact that we have this curiosity that there actually is a reason for everything, and the universe might have no cause. Rowe suggests that "the fact ... that all of us *presuppose* that every existing being and every positive fact has an explanation does not imply that no being exists, and no positive fact obtains, without an explanation. Nature is not bound to satisfy our presuppositions."[3] Others have suggested that the universe does not need a cause, or that it may be self-causing, or that it could have an eternal past (and if so, it would not need a first cause).[4] I am suggesting that none of these alternatives is acceptable from a logical point of view. They are not rational to any fair-minded person, and the last suggestion goes very much against current scientific evidence for the big bang. The claim that the first event (say the big bang) could have caused itself is not reasonable given all we know about physical events, knowledge (let us not forget) that is used as the *basis* for the conclusion of naturalism in the first argument above. These suggestions are simply grasping at straws when faced with what has looked to the vast majority of people very much like an insuperable logical difficulty. To put the point in another way, if one is going to offer a theory that explains all of physical reality, it has to explain also the *origin* of physical reality. Otherwise, one has left out the most important question of all. It is little wonder that naturalists frequently downplay this huge gap in their argument, and it is not unusual to find prominent naturalists today who are supposed to be offering us the latest sophisticated theories of reality based on reason and science ignoring this central question about reality, or papering over very difficult logical problems for their views. It is fair to ask why would studies that are supposed to be addressing the most basic questions of the universe and of life ignore or downplay one of the main questions?[5]

Other answers sometimes offered by naturalists will not do the job either, such as the claim that ours might be just one universe among many (a theory that has no scientific evidence to support it); the problem with this answer is that it does not address the ultimate question, it only pushes it back further because we would still need to know how, on a naturalistic hypothesis about reality, these multi-universes initially came into existence. One cannot explain (ultimately) the existence of an apple by saying that there were initially ten apples, and that one of the more recent apples gave rise to the present apple! This is a local explanation for the present apple, but it is not an ultimate explanation.[6]

It is this line of reasoning when developed to its logical conclusion that is the basis for one of the strongest versions of the famous cosmological argument for the existence of God. Although my aim is not to defend the view that God exists, a conclusion I am assuming throughout this book, but rather to show that naturalism on its own terms cannot be true, it is instructive to end this section with a few brief reflections on the implications of our above arguments for the positive case for God's existence. I would like to use an analogy of two goldfish in a bowl to explore this question (the "goldfish cosmological argument"!). One goldfish favors a theistic account of reality and the other is a naturalist! The theistic goldfish developed his view initially based on his realization that "there has to be a God; otherwise who changes the water?"! He reasons from this point to the conclusion that there must be a cause for his own existence, for the existence of the water, for the bowl, and so forth. The naturalist goldfish thinks that the bowl caused itself, though he freely admits that events *in* the bowl don't cause themselves, for example if the fish wait too long to eat their food, the water dissolves it (and so the evidence from a study of causes and effects in the bowl is firmly against any physical thing causing itself). He sometimes speculates that the goldfish bowl was caused by another goldfish bowl; at other times he argues, after he has been reading the work of a few philosophers, that although events in the goldfish bowl clearly have causes, the bowl itself has no cause; it just exists. My argument in this section is that the position(s) of the naturalist goldfish logically cannot be true. No matter what one thinks of the position of the theistic goldfish, we can be as certain as we can be about anything that the naturalistic position is not true. It is not reasonable to think that the bowl caused itself, that it just exists and has no cause, or that if it was caused by another bowl (another physical cause) this would be an ultimate explanation for the bowl. It is the same when we apply this reasoning to the universe itself.

But this line of reasoning does prompt us to ask: given that the naturalistic view is false, which view is true? And this is where the cosmological argument comes in, especially that version of it developed by St. Thomas Aquinas, and others.[7] A key further move in the argument is to reason that the cause of the universe (or of the goldfish bowl) must be *outside* of the universe. Again, we need to remind ourselves that we are seeking the ultimate cause, not the local cause. If it turned out that some other physical event in another universe caused our universe, we would want to know what caused this event, and so forth. St. Thomas's reasoning is that the question about the ultimate cause leads to only two possibilities. The first is that there must be some kind of cause that itself needs no cause, and this would have to be an eternal cause. The second is that there cannot be such a cause, and so then there is no answer to the question, and so the universe is ultimately unintelligible.

St. Thomas based his reasoning on an appeal to two interesting metaphysical notions, contingent being and necessary being. A contingent being, or a contingent event, is a being or an event that is not the cause of itself. A necessary being is a being that has always existed, and so does not need a cause. A series of contingent events linked together by cause and effect is called a contingent series. St. Thomas holds that the universe is made up of individual events none of which is the cause of itself, and so the universe would be a contingent series (just like the goldfish bowl, with its fish, water, vegetation, terrain, etc.). But a series of events like this requires an ultimate explanation for its existence, *no matter how many members are in the series*. We can explain the local cause of any particular event, or sequence of events, by pointing to prior causes in the series, but this will not, St. Thomas argues, help us to explain the existence of the *whole series*. So it is therefore reasonable to conclude that there must be a necessary being. Otherwise, no contingent series can get started. And we know that the contingent series in our universe *must get started somehow*! If we keep going back in terms of prior causes to explain any event, we have an infinite regress on our hands (as we have noted above), and can never give a full explanation for the present event we are trying to explain.

It is important to recognize that the existence of a necessary being is not a postulate of the argument, nor does St. Thomas approach the argument by assuming the existence of necessary and contingent being at the beginning. He begins with the existence of contingent being, or contingent events, and reasons backwards as it were logically to the conclusion that there *must* be a necessary being, no matter how unusual this sounds, or how difficult it is for us to grasp such a concept. There must be a necessary being because logically there are only two possible answers to the question of how did a series of contingent beings (the universe) get here. The first is to argue that it was brought about by a contingent being or cause, or series of causes. But this will not do because we can always ask what caused this contingent being, and so forth. The second is to argue that there must be a necessary being who started off all contingent events, and who does not himself need a cause. Otherwise, we would have no way to explain the existence of the universe. Of course, we could always refuse to pursue this line of reasoning, and simply conclude that the answer to our question is that the universe has no ultimate cause, has no explanation, that it is just here (that it came uncaused out of nothing, à la Atkins and Hawking). But this explanation is simply not acceptable because it is not reasonable. St. Thomas is asking: is it reasonable to think that our universe came into existence without a cause?

If we are to be guided by reason, we must conclude that no matter what the first event is, it must have a cause. Otherwise, we would have to speculate along with the naturalistic philosophers mentioned above that it has no cause at all; that it came into existence without a cause, just appeared, as it were. But this goes against all of our reason and experience with empirical evidence, including, crucially, all of our work in science. As Dallas Willard has noted, "even if it were neither self-contradictory nor counter-intuitive to suppose that something originated without a cause, the probability of it relative to our data would be exactly zero. There is … not a single case of a physical state or event being observed or otherwise known to originate 'from nothing.'"[8] This argument points to a necessary intelligence behind the universe, rather than to a beginning that has no cause, an infinite past that has no cause, or either a beginning or infinite past that have only a contingent cause.

Although we cannot go into this intriguing argument any further here, and there is a (very) large literature on the subject, supporters of the argument go on to argue on purely philosophical grounds that the necessary being is a personal being (rather than some impersonal force, which would create problems of its own), and is all-powerful, all-knowing, and all-good, the traditional attributes of God. But my aim here has been to show that, even if one does not pursue the cosmological argument further, or questions how far we can go with such an argument, the key point is that the naturalistic answer to the question of ultimate origins is not reasonable to believe. Whatever explanation of the universe and of human life turns out to be true, it cannot be a naturalistic one; this is because the universe (physical reality) must have a cause, and so naturalism cannot be true. I am reminded of the joke about a naturalist who informed God that there was now no need for him because everything could be explained (and indeed produced) by physical causes from scratch working with matter and energy, including life itself. God asked the naturalist if it was really true that human beings could produce life from scratch. The naturalist assured him that it was. To put the theory to the test, God proposed that they individually create a living thing. The naturalist agreed, and reached down for a handful of dirt to begin the experiment. But God stopped him, saying, "Not so fast. You said you could produce a life from scratch, so get your own dirt"! A perceptive joke, that raises a powerful question that we cannot ignore.

Do we come from stardust?

In pursuing our discussion of the implications of my arguments in this book for matters in religion, I would like to turn to a well-known remark of astronomer Carl Sagan's, a remark that many secularists latch on to when they are trying to suggest that there is no God. Sagan noted in his Television series, *Cosmos* (1980), that "our planet, our society, and we ourselves are built of star stuff."[9] Since then the phrase has caught on, especially among atheistic secularists, though it is more common now to refer to "star dust" instead of "star stuff." The phrase could be read in a number of ways, but its main connotation seems to be that we are *no more* than stardust, that we originated by a series of chance events, and this is meant to diminish our status, and to suggest also that there is no design behind the universe. I believe that this view of Sagan's is wishful thinking, that it is irrational and implausible, that his view is one that no reasonable person could believe.

In the first section, we looked at the question of the ultimate origins of the universe, and now we must turn to look specifically at *our* ultimate origin. For Sagan's famous remark is based on applying the idea of evolution to the universe itself—in the notion of cosmic evolution. He then extrapolates from the notion of cosmic evolution to the origin of life and of *Homo sapiens*. Cosmic evolution, as we noted in Chapter 4, is the idea that the universe began in a certain way, went through various stages by means of scientific laws operating on the initial ingredients, which produced the early atoms and other elements, and then eventually led to our present planets, stars, and galaxies, including earth, in their present states. Sagan's remark is also crucially meant to suggest the *unguided* nature of evolution (both cosmic and biological), to suggest that it all happens by chance (in the way described in Chapter 5), as well as to undermine any argument for an intelligent designer, or a God who is responsible for the overall outcomes of evolution. To say that we come from stardust is also a way of expressing Sagan's contempt for religion. (Of course, if God exists, he could bring creation about in the way Sagan described if he wished.) But what Sagan is suggesting with this remark is that the earth formed out of the debris of the big bang after a sufficient interval of time (about 5 billion years ago). Then about 3.5 billion years ago life originated on earth out of non-life, initially as a single-celled organism (this is not a respected scientific theory but a speculative hypothesis, but, as noted also in Chapter 4, it is a further way of extending the "mythic power" of evolution to other difficult questions for which Sagan needs answers).[10] The process of evolution then got started, and here we all are, with *Homo sapiens* at the top of the evolutionary tree at least in terms of sophisticated biological development, higher faculties and achievements (if not necessarily in terms of moral worth, according to some who propound this theory). So, according to this view, the origin of the causal chains that led to the arrival of *Homo sapiens* can be traced back to "star stuff" or "stardust" (and therefore, they usually add, not to a creator, though this conclusion is obviously a non-sequitur, a point to which I shall return in the next section). In short, this view is meant to undermine the notion of a creator, to remind us both that the processes of cosmic and biological evolution really occurred, and that they are governed significantly by chance; moreover, we are at the top of the evolutionary tree by accident, not design, and therefore differ only in degree and not in kind from other species.

Applying the arguments of our previous two chapters to this reasoning, we can see that it is incorrect in many respects. This is because supposing we grant to Sagan the whole story he tells, which is quite radical and far-fetched, and hard to believe without good, detailed evidence at each step, the conclusions he draws do not follow. The key point we need to emphasize again is that there is no chance involved in the process that Sagan is describing, a fact that Sagan is well aware of if he only focused more carefully on how science works, and a fact he relies upon to do his own work in science (in the way explained in the previous chapter). Sagan simply could not do astronomy if he thought that there were significant elements of chance involved in the way that causes and effects operate in relation to stars, planets, and galaxies. But if there is no chance involved in the cause and effect process, then, given the initial conditions—the big bang—our universe *had* to form just as it did, earth *had* to come about in just the way that it did, and life would *have* to emerge at the time and in the way that it did, the process of evolution would *have* to go the way it did, and man would have to emerge at the top of the tree, with our specific properties (consciousness, logic, reasoning, and moral agency). No other end result is possible, given the starting conditions (unless God or some other *free* agent intervened directly at some point or points during the cosmic and biological processes). Sagan thinks that this whole process all happened by chance because, like many of the thinkers we have discussed in earlier chapters, he thinks that the causes that brought about the effects in physics that he and all scientists study could have been different than they in fact were. This means that a different group of stars, for instance, could have formed than the ones that actually did form. Different galaxies might have arisen; we might have had a different earth, or no earth at all, and so forth; even different atoms and elements might have emerged from the big bang (why not, if prior causes are not necessarily determined by their prior causes, and so causal effects do not come about by necessity?). Sagan also believes that the outcomes of evolution are random because the causes of mutations could have been different, and so too could the causes that produce environmental conditions in which species evolve. I have argued earlier that none of this can be the case.

The whole discussion about chance and design in the universe, and in evolution, about necessity in nature, and about genuine or real chance versus probability, has been completely hampered by these misunderstandings on the part of leading thinkers in the field. This confusion about the role of chance in the universe is found in countless thinkers across the whole spectrum in philosophy, theology, and science.[11] But if I am right, it means that the emergence of the human species is not random but had to emerge from the process of evolution, and so this leaves completely open the question of a designer. It also means that we are not entitled to conclude from the claim that we have come from stardust that there is no designer, that we did not have to exist, that we are not special. All of these issues are still with us, even if all of Sagan's claims are correct, and that in itself is far from established. All of this is quite in addition to the key question that we have not yet considered (but will do so later is this chapter) as to whether God could introduce genuine chance into the universe and yet *still* direct the eventual outcomes (and how he might do this). But the main point is that there is no element of chance in the process Sagan and other thinkers are trying to describe. Given the initial conditions, and a deterministic universe, our planet earth and our species had to come about. These are facts of the most enormous significance. They force us to ask the question we addressed in the first section of this chapter concerning how did the whole process get started, a question to which Sagan himself does not give sufficient attention.

Sagan admits that the origin of the universe is "the greatest mystery we know" but does not pursue the question much further.[12] In a later book, he appeals to some of the arguments we have already critiqued above, opining that "an infinitely old universe has no need to be created. It was always here. If, on the other hand, there is insufficient matter to reverse the expansion [of the universe], then this would be consistent with a Universe created from nothing."[13] Sagan believes we will find the answer in time through science. There is a certain irrationalism to this overall position, especially when one considers the radical nature of its claims, and the inability of its proponents to back them up with arguments and evidence, their inability to recognize the limits to the scientific method. Indeed, perhaps this view is best understood as a "narrative" that one lives by, an *attitude* that one adopts toward life, a kind of commitment or faith in the power of science, rather than as a philosophically developed worldview. However, we must ask the obvious logical question of Sagan, Hawking, and those others who insist that this line of reasoning is plausible: is it reasonable to believe that the universe came out of nothing, if there was no prior cause or agency involved? Is it reasonable to believe that there is a *scientific* answer in terms of physical causes and effects to the question of the *ultimate* explanation of the universe?

Divine action in the universe

It is time to think about how God might have created the universe, with special regard to the question of evolution. Overall, I am saying that God created the initial ingredients of the universe (and we will assume for the purposes of this discussion that the big bang theory of how the universe came about is broadly true), and that he also created the laws of physics. As we have seen in our earlier analysis, these laws express the various relationships between the ingredients that make up the universe; that is to say, they are our expression of how the ingredients of the universe *behave* during their myriad interactions with each other. This means that the matter and energy and other ingredients out of which physical reality is made behave in certain ways; we discover these ways progressively in scientific investigation, and we call them the "laws of physics" (or the "laws of science," "physical laws" or sometimes even the "laws of the universe"). We do not invent these laws, or project them onto reality; they are already true, and we discover them. We can then utilize them in the future to understand and manipulate reality at least in some ways to our advantage because we have free will. If we did not have free will, we would be just another cog in the machine, as it were. So God created the ingredients out of which the universe is made, and gave these ingredients a nature that governs the way they behave when they interact with each other in myriad causal chains that produce effects, big and small, in the universe (as illustrated in previous chapters). So when Newton studied, and performed experiments on, the motion of various forces and their application to physical objects, he discovered that lawful relationships obtained, one of which was that "for every reaction there is an equal and opposite reaction" (his third law of motion). At the more simple level of ordinary life, a person might discover that if he closes the door of his car with too much force, the weather stripping begins to fail. Although, he does not think of it this way, the laws of physics cause this effect. But he knows that over time he can prevent the weather stripping from failing, or at least slow down its failure, if he is careful to close the door each time with less force. In doing this action, he is assuming the consistency of the physical laws involved in the process.

I have also been arguing that there is no chance involved in the universe, no random outcomes that occur in causal chains; that the universe is, in fact, completely deterministic. What are the implications of such a view for our understanding of God's intentions with regard to nature and species, including us, with how God might act in the world, and for other matters in philosophy of religion, theology, and morality? Let us first consider the development of the universe before life appeared. It follows from my view that God created the initial ingredients and the laws by which they behave, and that each subsequent stage of the universe emerged in a deterministic way out of the previous stages, in the way described earlier. This process continues up to the emergence of life. The whole process would also lead deterministically to the arrival of those ingredients on the scene that are necessary for the emergence of life (just assuming for the sake of discussion that life emerged from non-life at a certain point in the development of the universe, say 3.5 billion years ago on earth). So given the initial conditions and the laws of science, which were created by God, life would have to emerge when it did. In short, its appearance and subsequent development would be intended, or designed; God would have set up the universe with built-in teleological goals so that life would appear at a certain stage. (This is like a machine, such as the Mars Rover, acting in the way that the person who designed it intended it to act—no other outcome is possible, given a deterministic universe.) In this sense, a religious view of reality is completely compatible with evolution; evolution would emerge out of the ingredients of the universe as they interact with each other and develop, and it would be God's way of bringing about living things, since God is the designer of the initial blueprint that led to all subsequent developments in the universe.

It is important to point out that these conclusions assume that God did not later *directly* intervene and change a causal process or set of processes as they were occurring, and so alter the course of events (in the way *we* are able to do after we come along!). Should we make the assumption that God designed and initiated the universe and then never gets involved afterwards (a reminder that we are talking about before the appearance of life)? It is not clear that we can make this assumption. After all, we know that God has the power to influence causal chains as they are occurring, if he wishes. For instance, on the view of God that most theists hold God would have the power to intervene in nature to prevent a storm, to cause a reaction that is twice the force of the initial action (and so break Newton's third law), to stop the earth spinning for a moment, and any other outcome he wished to bring about (perhaps constrained only by the laws of logic).

The view that God gets involved in causal chains *after creation* is one that nowadays does not find much favor with either theologians or scientists. What reasons are advanced to support this view? One often hears theologians, such as John Polkinghorne, Arthur Peacock, Nancey Murphy, Philip Clayton, and others, suggesting that God does not "interfere" in his creation after he designs and initiates it.[14] The word "interfere" is a loaded term, and my view is that it should not be used in this case; because of its negative connotation it often seems to be aimed at carrying the conclusion without supporting it with any clear reasons. When one says that God would not "interfere," one is suggesting that there is something wrong with doing so, but I must confess that I cannot see what would be wrong with God changing causal chains if he has a good reason to do so. God's intervention in nature is often offered also as the definition of a miracle. Sometimes a miracle is described as a "violation" of the laws of nature. But again the word "violation" should perhaps be used with caution—it suggests without any argument that the laws of nature should never be, or perhaps even cannot be, changed by God at any time, something that is clearly false. Now, if one argued that God *would* not (not that he could not) "intervene" in nature, one must give a reason why. Although these are very difficult areas to think through, and the best we can hope for perhaps is intelligent speculation, it is not easy to see what the reason might be (other than that a particular theologian might not like the idea of God interfering in nature, might think that God should keep his "hands off" his creation—a very modern, individualistic view indeed, and one that is bound to be abandoned in future generations!). Describing God's becoming involved directly in his creation at a particular time as a "violation," or as an "interference," sounds very like a personal bias against the idea, rather than a substantive argument.

One reason sometimes given for why God would not "interfere" is that this might sound as if one is saying that God interferes to correct an imperfection, a view that would not be acceptable to hold concerning God's creation. This is because it suggests that God did not create the universe correctly in the first place, and this would be incompatible with his nature as a perfect, all-knowing, all-powerful being, and so forth. One might wonder why God would need to get involved in creation once he has set it up, and preordained certain ends? I am not sure how fruitful it would be to speculate about an answer to this question, but I would say negatively that I don't think that the only reason God might get involved is to rule out an imperfection. It seems that the type of creative activity that God might use is so far beyond what we know that we must keep an open mind on whether or not he might get involved in creation after he has initiated the process. Some theologians think that God "intervening" in nature would be a violation of our free will (after we come along), but again this seems too strong.[15] God could intervene from time to time without compromising human free will to any significant extent, just as human actions affect the freedom of others, but do not prevent the exercise of reason and free will. The view that our free will in the universe has to be a *total* free will is very hard to sustain, especially when one considers the constraints of the existence and actions of other people, as well as of the natural world. A few more constraints, or paths of guidance perhaps, introduced directly by God (especially since he introduced all of the others) is unlikely to compromise our integrity as rational, free beings.

Another argument against the conclusion that God intervenes in his creation might be offered from the side of science. One might argue that the application of the scientific method and the ever-increasing range of scientific knowledge shows that all events have causes and that there is no outside interference, no outside happenings that alter events beyond how their behavior can be explained in terms of cause and effect, as expressed in the scientific laws of the universe. That the scientific evidence, in short, shows that there is no outside "interference." This argument is a probabilistic one, since we cannot examine every cause and effect in history, nor can we know the complete causal explanation for most events. But the argument is that for the ones we do know about, we find that the nature of the physical ingredients, together with the relevant scientific laws, are enough to explain them completely, and there is no evidence of an outside cause or designer involved at any point in the process.

There is another important point about this particular argument of course, that some supporters of it might forget and that might be inconsistent with other views they hold. It assumes a *deterministic* universe. For if one allows into the universe elements of chance and randomness, then this means that a cause would not necessarily lead to its effect, and this would leave a space that would require explanation for why the effect came about. This would then undermine the deterministic assumption that is the underlying rational foundation for the probabilistic argument, and so the probabilistic argument is itself undermined. (Let us remember that we are not just talking about prediction here; we may not be able to predict which event will happen with accuracy given a certain cause or causal chain, but what this argument is saying is that for every event that did happen, there is a chain of causes that led to that event (and so no outside "interference" is necessary), and this conclusion *requires* a deterministic view of the universe. Those who argue against the possibility of miracles usually offer an argument of this sort: that when we investigate any alleged case of a miracle it will turn out to have natural causes, and that, *given* the natural causes, the event *had* to happen.) I believe that this probabilistic argument is a very influential argument, especially among a certain group of modern theologians (some of whom are scientists as well, like Polkinghorne and George Ellis).[16] These thinkers are perhaps intimidated by modern science, and allow it to influence their theology more than they should.

There are three other problems with saying that modern science seems to show that God does not intervene from time to time in some direct ways in history. The first is that in general it seems inconsistent with what theists believe about God. As noted, theists generally hold that God created the universe and has the power to intervene. Theists believe that in principle

God can intervene in creation, and since it is also an important belief of many theistic religions that God is immanent in creation, it then seems reasonable to think that God might intervene in nature occasionally. The second point is that God could intervene in nature without leaving any traces of his intervention. This would mean that our scientific investigations would never reveal any evidence of God's intervention in nature.[17] We cannot rule out this possibility. If we reply that if God intervened in nature, he would definitely leave traces, we need to explain why. Although these matters are necessarily speculative (in addition to being extremely interesting), I don't see why God would have to leave traces, though from our point of view perhaps it might be better if he did. One might reply that if God did not leave traces of his action in history, we are not entitled to believe that he ever intervened, because we need positive evidence on this matter, and the absence of evidence is not good enough. But this type of argument does not convince me in this case. This is because we already hold that God exists and created the universe, so given this, then it is reasonable to think that God might be intervening from time to time in creation, even if there are no traces of his having done so. We are asking: given that God exists and could intervene, did he do so, and would he leave any traces of having done so? I am not sure one way or the other on this question. I don't see why it would be bad for God to intervene, or that if he intervened, he would leave evidence of having done so.

The third reason for why one should not be too quick to say that modern science has shown that God does not intervene directly from time to time in his creation is that, in fact, modern science does not show this! This is because there are countless cases of alleged miracles, and other mysterious unexplained events. To argue that they all have naturalistic explanations is a promissory note rather than an established fact. It is also true that, concerning past events that occurred thousands, millions, and billions of years ago, we will never know the causal chains in enough detail to claim with any confidence that they show there was no intervention by God in the processes. One might rule out any possibility of intervention by God from time to time in the natural universe because of one's *naturalistic commitments*, but this would be a question-begging move, of course. One needs to be careful, however, that naturalistic assumptions, or too much deference to a naturalistic attitude, are not preventing one from having an open mind on some of these interesting questions about God's action in the world. This also means that the biblical miracles, such as the Resurrection, must be taken seriously, rather than simply interpreted as metaphors, or being ruled out by definition (as a matter of principle, as it were), because of intimidation by modern science.

One might wonder if it follows from the view that I have argued for so far, that God then *intended* for everything to happen in the universe that has happened. This would seem to be a consequence of the view that the universe is deterministic. Like a car engine, the universe gives the output it is supposed to give, that which is intended by its designer. This question raises two issues in particular, the problem of evil, and the issue of God's intentions. With regard to the problem of evil, the question is: if the universe is deterministic, and there are no truly random outcomes, does it not follow that God intended for all the evil events to occur that have occurred, and would this not only be inconsistent with God's plans, but also an argument against God's existence? One might press the argument a bit further by saying that the theory of evolution shows us that evil is in fact far worse than we first thought because it seems to be at the very heart of nature, and involves many seemingly pointless acts of violence, suffering, and death that are built into nature itself. These are all very legitimate observations, and I will leave a full discussion of them until our last chapter. With regard to the question about God's intentions, it is not entirely clear that if we say that the universe is deterministic, that we have to say that God intended all of its outcomes, though this does seem to be the most likely conclusion. The issue is complicated somewhat by the appearance of *Homo sapiens* with free will because this means that we are free to alter causal chains from our appearance onwards, and this interference by us can have a dramatic effect on future events and outcomes. Nevertheless, this would not apply to events in the universe before our appearance, nor would it excuse the physical constraints that the universe places even on human free will. These considerations really return us to the problem of natural and moral evil, which we will leave until the next chapter, but our conclusion for now must be that if the universe is deterministic, it would seem to follow that God intended all of its happenings and outcomes, the good as well as the bad, at least until our appearance.

What is God's role in evolution?

We must now come to the time on earth when life appears, leading eventually to the appearance of *Homo sapiens*, and to consider especially God's role in the process. Let us recall our assumptions and arguments so far. We are assuming in this book that God exists and that evolution occurred, and I have argued at length that the universe, including the process of evolution, is deterministic. Given all of this, it would follow that God brought about all of the various species by means of the process of evolution, that this is the way

God intended for the species to come about (and so there are no "random" species). In this section, I would like to try to develop some of the interesting implications of such a view, and this will also give us a chance to address some of the implications (discussed in Chapter 4) that have been claimed for the theory of evolution.

Thinkers in traditional philosophy and theology, under the influence of St. Thomas Aquinas's analysis of causation, often appealed to the distinction between primary and secondary causes as a way of explaining how God acts in his creation.[18] They argued that God is the primary cause of the universe and life because he created it out of a set of initial ingredients, together with the laws of science, and built in certain ends at the beginning of the process. Secondary causation then refers to the fact that everyday physical events in the universe (including in evolution) are governed by scientific laws. Our work in science deals with this domain of secondary causation (and many today would insist that science should be *defined* as the study of only secondary causes). Proponents of the distinction between primary and secondary causation also usually hold that God does not *normally* intervene in the day-to-day operation of physical events (the secondary causes). However, God can perform miracles whenever he wishes, and often does so, especially by responding to prayer, and God is also immanent in creation in several ways.[19] So this view is not a form of deism because, although it proposes that God set up the laws of nature as a way of guiding the universe to unfold in a certain way, it does not rule out God's intervention in nature from time to time, as the Deists generally did. St. Thomas also held that God is continually sustaining the universe in existence (including its in-built teleological goals) because it is not capable of doing this on its own. Human beings can also intervene in the process of secondary causation through free will, which is what science is all about, since the discipline is not just concerned with understanding the universe and acquiring knowledge for its own sake, but also wishes to utilize this knowledge mostly for the benefit of mankind by interfering with the (deterministic process) of secondary causation (for example, by preventing a disease from taking hold in humans, or by protecting a crop from insect damage). While the notions of primary and secondary causation are very helpful, and indeed almost indispensable in thinking about the nature of the universe, and the difference between local causes and ultimate causes, the distinction still does not address in terms of specifics the way in which God operates through evolution. The distinction does not tell us why God might use evolution, and if he did use it, how it works in terms of cause and effect, chance and determinism. We need still to try to elaborate further on these questions.

Is there anything wrong with saying that God created the species, and especially *Homo sapiens*, by means of evolution? And if he did so, how does the process work in terms of specifics? (Leaving aside the point about the problem of evil, which I will come back to in the last chapter.) Many evangelical scholars argue that one major problem facing an evolutionary account of creation is that it conflicts with the Bible, especially the account of creation given in the Book of Genesis. As I have argued elsewhere, and won't repeat at any length here, I favor St. Augustine's (and St. Thomas's) approach to dealing with this objection.[20] Writing 1,400 years before the theory of evolution was first proposed, St. Augustine argued that the Book of Genesis is making three important points about creation: that God created the universe according to a plan; that God created the species; and that man is at the top of creation, or what we would now call the evolutionary tree, by design. Augustine allowed that the biblical writers might have used occasional literary license to describe *how* God brought about these events, but *that* he brought them about was the key point. Nor does this conclusion undermine the Bible as a revealed text because we can generally distinguish between which passages might be given a more metaphorical reading, yet contain deeper points that are not revisable, and those passages that are intended to describe real events. So on this understanding Augustine thought the creation story may be an example of the former, and the Incarnation and the Resurrection narratives are examples of the latter. So in general this type of reading of the Book of Genesis does not undermine the Bible.

Augustine held that when religion and scientific conclusions seem to conflict, we need to look at both sets of claims carefully, and not worry about real conflict, for there is none. We will never discover a scientific theory that will truly undermine religion. Although a consistent theory of biblical interpretation can be hard to develop, I think that St. Augustine's approach is a very fruitful one. It allows us at once to be informed by scientific knowledge but also to work out how this knowledge coheres with our theology, and with our approach to, and general understanding of, the more ultimate questions. We do not need to adopt a defensive posture about a possible conflict between science and theology, for there is no conflict, and apparent surface difficulties can be resolved quite well. This general approach is a version of the view that is today known as theistic evolution; it has been expressed by Pope John Paul II, and has recently been repeated by Cardinal Christoph Schőnborn.[21]

However, one might still argue that if God created the species by means of evolution, this means that man differs only in degree from other species, thereby undermining the status of man in creation. There are five points to make in response to this objection. The first is that the argument that man differs only in degree and not in kind from other species is usually advanced by naturalists; it is not one many theists accept. This is because the evidence that human beings differ in kind from other animals is overwhelming. This consists not just of man's achievements, but also of our remarkable human qualities, especially consciousness, self-awareness, reason, logic, and free will, and other related features of human life, knowledge and experience, such as the nature of truth,[22] the coherence of mathematics with reality, our pursuit of objective knowledge (including in the area of morality), and experiences like love, compassion, justice, and so forth. As codiscoverer of the process of natural selection and contemporary of Darwin's, Alfred R. Wallace noted, "Natural selection could only have endowed savage man with a brain a little superior to an ape, whereas he actually possesses one little inferior to that of a philosopher"![23] Naturalistic attempts to explain these features of human life purely in terms of physical traits or as having completely naturalistic origins have run into insuperable problems, so much so that much work today both in philosophy and science simply *assumes* naturalism as a starting point, and goes on to speculate about what might be the case *if naturalism were true* (the philosophy of mind and language is dominated by such assumptions, as is much work in psychology), or simply ignores the problems altogether. (I discuss some of these qualities in more detail in the last chapter in our examination of morality and free will.)

The second point to make is that the fact that human beings both understand and can control to a significant extent evolution strikes me as being *an enormously significant point*, and is a further argument that we differ in kind from other species. We are now at the stage where we can control evolution to such an extent that one might be tempted to conclude that the process will essentially stop with us; that we will remain at the top of the tree of life because we will be able to prevent our ever becoming extinct (and so no higher species will evolve from us). Given our control over evolution, it is also very unlikely that the human race will ever be wiped out by a natural disaster or a disease. It is highly likely that the human species will never become extinct by the ordinary process of evolution on earth, and this is an argument that human beings are intended by God to be at the top of the tree, and have been given intentionally the ability to understand and to control the process of evolution. Our ability to control evolution also means that a species has now evolved that *can* direct mutations according to the needs of the organism, or to invoke Dawkins's and Sober's language from Chapter 5, a species has evolved that *can* detect which mutations would be beneficial to its survival and development, and can take steps to bring them about. *This is an incredibly significant development in the history of the universe*, and signals a clear difference in kind from other species.

A third reason that lends support to the view that man differs in kind from other species is the progressive nature of evolution (discussed in the previous chapter). Progressivism in evolution is also evidence, as Keith Ward has noted, of design in the universe.[24] Evolution is progressive in that it produces more and more complex species over time, culminating so far in the appearance of the most advanced species, *Homo sapiens*. This either happened by chance or by design, and it is not believable to say that it came about by chance. This is a *very persuasive* argument that the evolutionary process was set up to lead to man as the intended outcome, and this is a further reason to think that man is in a different category. (Yet we cannot rule out that the future of evolution might lead to the arrival of a species that is even more advanced than us, that *this species* might be the intended outcome, however unlikely our understanding of, and control over, the process makes this possibility.) A fourth reason to support the specialness of man is provided by the biblical events of the Incarnation and Resurrection. One can argue that the fact of Jesus taking on the body of a man, and of then been raised from the dead, shows that human beings are the intended outcome of creation, and that our salvation is the main plan of God for creation, the main reason behind creation. God becoming man is obviously a very significant event in history, and it goes hand in hand with the biblical teaching that man is made in the image of God. So here we have a biblical reason for why man differs in kind from other species; it coheres with the biblical view very well.[25] And we might also consider, finally, the view that perhaps it does not matter so much whether or not we differ in kind from other species, at least in terms of categories. One might argue that the facts are all that is important, including the fact that we have enormously significant properties, including knowledge of, and control over, the process of evolution; these facts distinguish us clearly from all other species that ever lived (and perhaps from all others that will live in the future). It is because of our different properties that we think we are in a different category from other species, but whether we are or not, it is the properties and not the category that give us a very special status in creation.

We might wonder how more precisely God created by means of evolution? I have been arguing that God designed the universe in a certain way to bring about certain goals, of which a main one is the emergence of man. God employed the process of evolution as the means of doing this. But how did he use the process? One thing I have argued for is that God created a deterministic universe, one that follows laws consistently, and so evolution would also follow a predetermined path. This means there is no chance involved in the process; the species that emerged therefore had to emerge and are not random. It also means that an important claim of many evolutionary biologists and one frequently repeated in biology textbooks, and in books on science and religion, asserting the role of chance and denying any role for teleology in the process of evolution, is false. But we are interested in the question, concerning divine action in the world, whether God intervened in any direct way during the process of evolution, or did he just set it all up in a kind of deistic fashion, with a preordained program built into it, allowing creation to evolve to its eventual outcomes? Some creationists and supporters of Intelligent Design theory have argued that God might (and probably did) intervene directly at certain points along the way, especially perhaps to bring about the creation of new species. Such interventions are sometimes described as acts of "special creation," and are often defended by arguing that the alternative argument based on an appeal to the thesis of natural selection proposed in the theory of evolution does not have enough evidence to support it.[26] Or by arguing, as Michael Behe does, that the cellular structures of much of biological life are so complex ("irreducibly complex," in Behe's phrase) that it looks as if they could not come about gradually by means of a process of natural selection.[27] Behe argues that it rather looks as if cellular systems were designed all at once as it were, and this at least suggests (though perhaps it does not require) an act of special creation (we will consider this view further in our last chapter).

I have suggested in the earlier chapters on evolution that the evidence that evolution produced complex cell structures and species entirely by means of natural selection is open to question, and is based on an extrapolation from the evidence from microevolution, an extrapolation for which it is reasonable to ask for more direct evidence. As I pointed out in Chapter 3, few would deny the extrapolation if the evidence for it was clear, but it is far from that. So I am sympathetic to any argument that speculates about alternatives as we grapple with the difficult task of trying to understand the origin of species, and God's role in their origin. But I think we must come to the same conclusion here that we came to above when discussing the development of the universe before evolution—that it is very hard for us to say one way or the other whether or not God intervened in the origin and development of life. From the point of view of science, the evidence is inconclusive. We have an interesting (and some would say powerful) theory that proposes that man developed from lower species through a natural process, but evidence in terms of specifics is rather sparse. (And the matter is made even more complicated by the fact that there are many who want *to reject* that this is the way it happened and many who want *to believe* that this is the way it happened, for reasons that are often not related to the question of evidence.) I find it very difficult to give reasons to support one view over another for how God created the species. It seems to me that the evidence that God created life and the species for a purpose, with man at the top of the tree, is very good, but it is harder to say whether he did it in a rather deistic way, or by some special interventions along the way. I don't think the scientific evidence helps us much here, at least not at present. But even if it became stronger so that there really was no question that natural selection could drive macroevolution and could bring about complex features of species, it would still not rule out God's special intervention in some way, as discussed above, though perhaps it would make it less likely.

Could chance be part of nature after all?

We can take this discussion further by considering whether God could allow some role for chance in the process of creation, but still bring about an intended outcome? This question is relevant because it is related to the issue of whether even if evolution did involve some element of chance, it could still be teleological in nature. In other words, even if evolution did contain an element of chance, would it follow therefore as many naturalists claim it does, that processes that led to the arrival of *Homo sapiens* must have been significantly governed by chance occurrences, and also that there is no design in nature, and likely no God? This is often the aim of such arguments about chance; the insistence on a significant role for chance that we see expressed with so much certainty in Monod's work is partly an ideological commitment motivated by his prior atheism. But could God allow for some chance in the universe, and in particular in the process of evolution? I am not inquiring about the possible role of chance in quantum theory, which I have already discussed in Chapter 6, but I am asking whether, in the ordinary process of causation in the universe primarily at the macro level, there could be an element of chance, and what its consequences might be for the questions of this book. I wish to explore this topic further by considering the thought-provoking views of two theologians.

Let us turn first to the views of John Polkinghorne, a theologian and physicist. Polkinghorne holds that there is an element of chance involved in evolution: "The fruitful history of an evolving cosmos has involved an interplay between [chance and necessity], not only in biological evolution on Earth but also in the physical development of the universe itself."[28] He calls this element of chance "historical contingency" several times, but never explains what he means by this term, nor does he give any examples of it. I think Polkinghorne means by it what Gould means; that is, that some causes, including in the process of evolution, *need not* have occurred, and so the events these causes led to need not have come about. History could, therefore, have been quite different; as Polkinghorne expresses it: "the specific structure we see is therefore present 'by chance' (it could easily have been different ...)."[29] One consequence of Polkinghorne's view is that there is no guarantee at the beginning of the universe which particular species will emerge from the process of evolution. Nevertheless, Polkinghorne thinks that God is directing a plan in an overall sense because the ingredients of life are built into the universe from the beginning (though he also seems to suggest that the initial ingredients might have been different). Unfortunately, Polkinghorne never elaborates on how this would work, but he thinks that the laws of nature would lead to the coming to be of self-conscious agents, yet these agents would not need to have the specific biology of *Homo sapiens* due to the presence of chance in the process. He is not clear about whether, if a different species of self-conscious agents had evolved instead of us, this species would have the distinctive characteristics we have, such as reason, free will, and moral agency (though he does think that some kind of conscious life would emerge). This view is interesting but it is too vague to be of much help, and by failing to try to tease out how the process would work, in the way I have attempted in the previous sections, Polkinghorne also invites the charge that his view runs a real problem of inconsistency.

The difficulty arises when we ask whether it is intelligible to say that God could allow for real chance in the universe and at the same time to still be directing the process toward a particular set of goals, such as the emergence of conscious life? On the surface at least it seems contradictory to say that there is some real chance—causes could have gone otherwise than they in fact did go, and so effects would be different—and yet, overall, the process can still bring about a particular outcome that has been designed in advance, built into the blueprint, as it were. Polkinghorne's view is even harder to make sense of because of his insistence that the laws of nature will lead to the coming to be of self-conscious agents, because this claim suggests that the process is deterministic, and so he seems to be advocating the presence of both chance and determinism in the process. Perhaps he could say in reply that in complex but localized causal chains in which the beginning leads eventually (say) to *Homo sapiens* that, *within this chain*, there could be chance elements. This would mean that some causes could be otherwise than they in fact were, or that many different outcomes are possible in the way evolutionary biologists often argue (the view I explored thoroughly in the previous two chapters and rejected). However, the end result of the causal chain might still be preordained, but, as Polkinghorne has noted, not in terms of specifics, but in general way, so that a conscious agent has to emerge, perhaps with certain properties like reason and free will, but not with a specific biological structure, or set of traits (or even essence?).

This seems to be Polkinghorne's view, but I do not see how it can work. For it seems contradictory for there to be a lawful process that then has elements of chance as well. This would mean that somehow, to recall one of our examples from Chapter 5, the seeds of the plant that poisoned the river that the beaver later drank from *might not* have landed on the river bank and taken root there. But for this not to happen the prior cause of this event would also not have to happen, and so on. As I pointed out, this is how physics works, a point we sometimes forget when we are doing biology. But if Polkinghorne were to make such a view work and be consistent, he would have to argue either that the laws of physics don't always hold, so that even though there is a prior cause that (lawfully) produces a subsequent effect, the subsequent effect does not always have to happen; or he could argue that God might intervene in certain cases in acts of special creation, as it were, and guide the process in some more direct way to bring about the outcome.

In the latter case, to the outside observer it could look as if there was an element of chance from the point of view of science, but there would be an in-built purpose all along guided by a designer. However, Polkinghorne rejects any kind of special creation or intervention, and firmly sides with that group of modern theologians who find this view very problematic, saying that "the universe is not God's puppet theater."[30] But again he does not explain how God acts in creation in any detail, and his rejection of God's intervening in some more direct way seems to be founded on the idea that he personally does not like this view, rather than on firm philosophical, theological, or scientific reasons. Polkinghorne notes that "the costliness and blind alleys of evolution are the necessary price to be paid for this open, exploratory creation," and that the universe is making itself.[31] The difficulty is that it is incumbent on him to say more precisely how this works; otherwise it is too vague to help us in our understanding of these matters. I agree with Polkinghorne that evolution is inadequate to explain complexity in nature in general, and in the human brain in particular. Polkinghorne takes this to suggest there must also be a designer involved, but he never really faces up to the specifics of how it all would work, and does not offer a convincing view of how chance could exist alongside causation in the universe, and yet still lead to specific (preordained?) outcomes.

To explore further the issue of whether the laws of the universe always hold as a way of thinking about a role for chance in creation, let us turn to the views of the theologian and biochemist, Arthur Peacocke, who has written an interesting, thorough, and very insightful book on the question of how God might act in the world. Peacocke begins his discussion by recognizing the clear distinction between how the universe obeys laws and acts in itself, and our ability to predict how it will act, a key distinction in my argument in Chapters 5 and 6. However, on the question of the uncertainty principle in quantum mechanics, which we discussed in Chapter 6, he appears to accept that some parts of reality at the subatomic level are indeterminate in themselves, and that this introduces some elements of chance into the universe.[32] There seems to be three levels where Peacocke thinks chance can enter into reality.[33] The first is that the behavior of reality itself at the subatomic level involves some elements of chance (and not just unpredictability); second, causal chains can interact with each other randomly, and this randomness can change the causal history of the universe, including in the process of evolution. Third, many effects emerge from within systems; these systems consist of their own structures, their environments and ingredients, the laws, all interacting with each other to produce effects, and *given the system* there is a range of possibilities that can come about in creation. This range is limited, according to Peacocke (citing the work of several other theorists who have written on this issue, including J. Maynard Smith and Karl Popper); so even though there is an element of chance involved, and we can't predict the outcomes of any system with complete accuracy, it is still the case that, because the system constrains what can emerge, the system can be said to have a goal, as it were. This is why we can broadly predict which effects might emerge from a system, even if we cannot predict the actual outcomes in terms of specifics. So even though there is an element of chance involved, there are also system and law constraints; so what we have is a combination of chance and law, and so the outcome could still be directed by God.

All of this is true as far as it goes, but I don't think Peacocke has really introduced chance into the universe with this argument. It is true that causes emerge from a type of system similar to the types he describes but it is still a fact on Peacocke's view that the existence and nature of the system itself (for example, a planetary environment or a species habitat) would be governed by chance (if chance is really present), and this would seem to preclude the system from arriving at any particular outcome or structure. So whatever outcomes emerge from the system would be significantly affected by the presence of chance happenings. Given this, it is quite difficult to then conclude that certain effects are *intended* to emerge from the system. Instead, it looks like the effects are truly random, and not part of a plan. This is the very reason that many naturalistic interpretations of evolution have such a problem explaining why the process leads to increasing complexity and progress in organisms like that of human beings, if, as Monod says, their emergence is significantly governed by chance. The problem with Peacocke's position is that, since each step right from the beginning of any causal process would be governed by a significant element of chance (in Gould's and Monod's sense), then it is very unlikely that we would ever get evolutionary progress and complexity of the type we have (this is why so many look at the nature of human life and doubt the current evolutionary account of it, especially the claims about randomness). If nature is moving toward complexity, this must mean that it was set up this way, which suggests a designer, and a predetermined set of goals—that is to say—a deterministic universe. One cannot have it both ways, it seems to me; either one says that there is an element of chance and the complexity is mostly an accident (which is not believable to any rational person, in my view), or one says that there is an intended outcome, and so no chance is involved.

Our second observation is that despite his good work in trying to keep the behavior of the universe in terms of causes and effects distinct from our ability to know about, or make predictions about, this behavior, Peacocke does sometimes still seem to confuse or conflate them. For example, when he says there is a range of possibilities that can emerge from a system, this is true but it seems to be only a point about predictability and not about how reality behaves. If we soak steel bars in a dish of various chemicals for a period of time, and then specify that there are only a range of possible effects that can occur, and identify them, this means only that we cannot predict the specific effects, but it does not mean that *any one of the range of effects actually could happen*. The same is true if we put ten goldfish of different varieties in a bowl with a view to them mating. Whatever baby goldfish are produced *had* to appear, even though all we can predict in advance is a range of possibilities (assuming that the goldfish don't have free will!). The fact that we could make our predictions more accurate, even certain, by reducing the number of goldfish in the bowl, or by becoming very knowledgeable about the specifics of goldfish reproduction, and by controlling for some of the other difficult-to-know variables, shows once again the deterministic nature of the physical universe. This is really the key point. As I argued in the previous chapters, the effects that happen have to happen because of the deterministic nature of the universe and the laws of physics. This means that there is no range of possibilities *in reality itself*, and so we are still left, it seems to me, with a deterministic universe. I think that in general Peacocke is right that there are limits on the changes that can emerge from any causal system, but it does not follow from this that the actual outcomes that do emerge could involve an element of chance.

Peacocke goes on to argue that the interplay of law and chance in the universe means that the universe is *creative*, and that we can say this is God's intention. Although his view is very interesting, and well developed, I don't find his argument convincing. Peacocke thinks that God has included some chance in the universe, as well as creating the laws of science, of course, so that the process of development, including in evolution, is more creative. He puts it this way:

> If all were governed by rigid law, a repetitive and uncreative order would prevail; if chance alone ruled, no forms, patterns or organizations would persist long enough for them to have any identity or real existence and the universe could never be a cosmos and susceptible to rational inquiry. It is the combination of the two that makes possible an ordered universe capable of developing within itself new modes of existence. *The interplay of chance and law is creative.*[34]

Peacocke's reasons for why the presence of chance would give us a more creative universe are not clear. I don't see how the universe would be repetitive (and therefore uncreative?) if it is deterministic. He probably means that if God preordains from the beginning what must emerge from the causal processes in the universe, that this would not be as creative as there being some random role in the process itself that contributes to the outcomes. The second alternative would mean that it is not foreordained in advance which species would come into existence (and perhaps it also entails that God might not have full control over the process), though Peacocke appears to think the system designed by God will constrain the outcomes so that some kind of advanced, conscious observers will prevail, though not necessarily *Homo sapiens*. He thinks this approach is more creative than God establishing a deterministic process at the beginning that arrives at predetermined outcomes. It is an interesting point, but I am not sure I see any significant difference in creativity. Surely the difference is only in the method used by God, rather than in the outcome?

Peacocke, however, believes that God may not know the outcome of the process of evolution because of the chance element that he has built into it. He does not know the specific outcome, but he would know the general outcome. This again is an intriguing idea. This is a limit that God has placed on his own omniscience, according to Peacocke, not an argument that God is not omniscient (just as we might say that he has placed a limit on his omnipotence by giving us free will). One overall problem I see with Peacocke's view is that it is not clear that if God has preordained the general outcome, that he would not then also have to logically preordain the steps to bring this outcome about, and so no chance could be involved in the process. These reflections prompt us to ask important questions once again as to whether it is really possible for there to be *genuine chance* involved in the causal processes that occur in the universe, and also about the place of design in the overall scheme of things. We will take up these questions, along with other significant ones raised earlier, in our final chapter.

8

Evolution and Design, and the Challenges of Evil and Morality

The relationship between religion and evolution is a fascinating topic, and our previous discussions have raised a number of vital issues that we should reflect on before we bring our analysis to a close. An interesting question raised in the previous chapter is whether, leaving to one side for the moment our thesis that the universe operates in a deterministic manner, the presence of real chance operating within causation (the view defended by Peacocke) would support the conclusion that there is likely no design in the universe, as many atheistic thinkers claim. This question prompts us to examine the universe more closely for some evidence of design. Coming back to our thesis of determinism, another obvious question raised by it is the problem of evil, as noted in the previous chapter. It would seem, if we hold to a deterministic view of how nature behaves, that, at least up until the appearance of *Homo sapiens*, any event that happens in the universe *had* to happen, including all of those events we would classify as evil, and this raises the question of why would God, who is normally conceived of as all-powerful and all-good, allow such evil? This chapter will consider these important questions. In our discussion of design, we will also explain how our view differs from Intelligent Design Theory. We will conclude with a reflection on the extraordinary features of man, especially with regard to the nature of morality and free will, both of which raise serious problems for naturalism, and suggest the presence of design at work in the natural process.

Design and chance

Our consideration of the thought-provoking views of Polkinghorne and Peacocke in the last chapter raises the important question as to whether it is possible for there to be real chance involved in the causal processes that occur in the universe. Although I did not find their particular answers convincing, their work raises this interesting question. Is it possible that at least some causes *could have been otherwise* and yet for there still to be a set of preordained end(s)? Many think this is intelligible.[1] I don't think it is—unless one posits special intervention by God from time to time to break the laws of physics, but, as we noted, Polkinghorne, Peacocke, and several other theologians all but rule out intervention by God in secondary causation. *Could* God have set up a process that is governed at least partly by chance but that leads to preordained outcomes? And if God could do this, *did* he set it up this way? We can perhaps examine scientific evidence to help us answer the latter question. Peacocke thinks that modern science shows some evidence of randomness. I have argued that this is by no means a settled interpretation of modern science, and also am inclined to the view that God could not (or would not?) set up a process with some element of real chance in it and still bring about a particular outcome, though I acknowledge that this is a difficult issue to make a judgment about. The reason I am inclined against this view is that I cannot see how an effect could come about in the universe if it has no prior cause, and if there is a prior cause, I cannot see how the effect could be otherwise than it is, and so forth. So the only possible way for an effect to occur by chance is for God to intervene and bring it about, but then this is not genuine chance; it is simply a different chain of causation, one influenced by an agent (and an intervention that may or may not be detectable by scientific investigation). This is where Peacocke and I disagree. Ours is not just a disagreement about whether the universe is deterministic in itself or not; we do disagree about that, but ours is also a disagreement about whether it is logically possible for there to be an element of chance in the way the universe unfolds (even within a system, like Peacocke advocates), and yet for the eventual outcomes to be preordained.

I hold that there is no chance because of determinism but also because it does not seem intelligible that there could be chance in the universe. I mean by this that I do not see how, given our understanding of modern science, causes could have been otherwise than they were (as explained in the previous chapters). I recognize that work at the quantum level has suggested to many the uncertainty principle, but there is considerable disagreement over how to interpret this, and I have suggested that the view that the universe itself acts randomly is not intelligible. One might argue instead that chance could enter the world at the macro level in this way: by saying that the laws of physics do not hold 100 percent of the time, but hold only *in most instances* (similar to the way some claim they operate at the quantum level). We would need to decide also whether this claim applies to all laws or only to some laws. Does it hold, for instance, for laws that are utilized in the running of car engines, such as electrical and chemical laws, as well as other general laws of physics? We need to distinguish two kinds of cases here to appreciate the point: (1) cases where a car does not start because of an unknown or hidden variable (for example, a leak in the fuel line affects fuel pressure and prevents the engine from starting)—this is a case where the laws do hold, and the unknown variable is the reason the expected effect does not occur; (2) a case where Ohm's law of electrical circuits did not hold (because laws do not hold 100 percent of the time), so the car failed to start because the electrical circuit in the fuel pump behaved differently in itself than the way it usually does (the car does not start in this second case because the law does not hold *in this instance*, not because of hidden variables, such as loose terminals, and broken wires). If there are cases of the second kind, one could argue that chance may enter into the universe in this way. Applying this thinking to evolution, the survival of a species could be affected by a disease that took root because *in one instance* a law of physics did not hold, and this could affect the course of evolution. So chance in these cases would just mean that the relevant scientific laws did not hold in a specific instance, and so the outcome that resulted is also a matter of chance. So, if these kinds of cases are possible, one might argue that the development and survival of species would be governed in a significant way by chance, as well as by laws.

I think that modern science is firmly against this argument for the presence of chance in the universe. Even though we often note that the laws of physics are statistical, we go on to extrapolate from our study of causation in particular cases that scientific laws hold in *every* case. This is our assumption in science, and its truth is why we can build technological machines that will work; it is also the reason we never consider that the laws may have broken down, however temporarily, when we are repairing machines, or, more generally, when seeking explanations for causal failure. So the statistical nature of the laws of physics is not sufficient to undermine our belief that each law has universal application; therefore, at the quantum level, when we see a fundamental problem with predictions, even in principle, we should interpret this to mean that our knowledge of the law may never be complete, but not that the law itself is not universal in principle. That is why we must be careful at the quantum level when we say that the particles appear to behave "randomly" some of the time. The behavior of the particles *only appears "random" to us* because we cannot fully predict their behavior even using our best measuring techniques, along with our knowledge of scientific law, but they are not random *in themselves*.

If it were true that the universe behaved randomly in itself to some extent, then some chance would enter the universe, and the outcomes would then be determined by the operation of laws and chance. The final outcomes would then be unpredictable, at least to us, and perhaps even an all-knowing mind like God's would not know when an element of chance would enter a causal process, and so would not know the end result of the causal process. This understanding of chance, where the laws of physics do not always hold, is consistent with my earlier definition of chance in Chapter 5 as meaning that the cause of an event could have been otherwise (and recall that our definition of randomness is that if the cause could have been otherwise, there would be no predetermined end result, and so the concepts of chance and randomness are inextricably bound up with each other). It is consistent with my definition because a cause that usually produces an effect might not produce the effect because the scientific law might fail in a particular case (because it holds only in *most* cases, not in every case). The seed might not have landed on the river bank because the law of gravity did not hold in that one instance, and so the plant would not have grown, and infected the river with its chemicals, and the beaver would not then have been poisoned, and his DNA would not have been altered, and so forth.

My view is that once we have a set of initial ingredients (remember that these ingredients include the environment) and a set of laws, the end result is predetermined. Unless an agent intervenes. By predetermined I mean that the final outcomes of main and subsidiary causal chains, with regard to local and ultimate outcomes, must happen given the starting point. This is why science works. We can't always predict which way causal processes will unfold but the way they do unfold is predetermined. Does this mean that the goals of the universe (the *telos*) are already laid out, and so there are no random outcomes? Yes, by whoever set up the initial ingredients. One can't do science if one does not make this assumption of determinism. For example, one could not describe, as physicist Steven Weinberg does, the formation and interaction of particles just after the big bang.[2] This is because Weinberg reasons backwards and forwards to and from the existence and interaction of particles though a chain of causes; these causes bring about certain outcomes because the laws of physics hold. But suppose Weinberg accepted that at many points along the way in the process that led to the formation of particles *some events happened by chance*, then his chain of reasoning would break down, and his account of what happened just after the big bang based on contemporary evidence, and the laws of science and of logic, would be fatally undermined. The evidence is strongly against the operation of chance in the universe in this way.

Peacocke is convinced that chance elements in natural selection could still lead to complex outcomes, and as support for this view he refers to Dawkins's example of the computer program he used to illustrate the concept of "biomorphs."[3] Basically, Dawkins developed a computer program designed to mimic the process of evolution and natural selection to some degree, where he started with simple patterns, and, after introducing random changes as part of the program, noticed how the patterns became more complex after many reiterations of the procedure. Dawkins was attempting to show that natural selection works in a cumulative way, and is not entirely random, since each later stage builds upon the earlier stages, and is therefore constrained by them. Dawkins argues that although the computer program has random elements, the end results are not entirely random because they are constrained by the system, and it is possible to predict in a general way what some of the outcomes will be. He also suggests that this type of computer program illustrates that complexity is likely to emerge relatively quickly.

Of course, he has shown none of these things. We need to take care and be very attentive to what is going on here. First of all, Dawkins and his programmers designed the system that they are using (including the so-called "random" steps in the program), so the fact that complexity emerges quickly (and indeed *every other effect* that occurs in the computer model) is *entirely* due to the initial design. Second, there is nothing random going on in the process. There is no such thing as a random computer program or, more accurately, a computer program that produces "random" results, just as there is no such thing as a random slot machine; there are only, from our point of view, unpredictable programs and machines due to their causal complexity. When we describe these processes as "random," or as involving random steps, this only means that we can't predict which combination the program or the slot machine will turn up, but the one that does come up happens entirely in a deterministic way. When a slot machine wheel spins, the position it stops at is *completely determined by the laws of physics.* The same is true of the causal sequences in computer programs. Peacocke, Dawkins, Monod, and others are arguing that there can be genuine chance in secondary causes. I think they have failed to pay sufficient attention to how modern science works. It is in fact very difficult to introduce a real chance element into nature from a conceptual point of view. This is because it would involve either arguing that the causes of individual events could have been otherwise, a position I believe to be impossible in our universe (excluding intervention by God or man), or arguing that, despite the very low probability of the existence of favorable conditions for life, including and especially complex life, there is no overall purpose or design in the universe, and this claim is very counter-intuitive and against most of the evidence we have.

The absence of chance in the universe does not mean, of course, that God does not intervene in the natural world, as we noted in the previous chapter. Indeed, our deterministic thesis does not commit us to any position on the *frequency* of God's intervention. It does not follow that because the universe, if left to itself, as it were, is deterministic, that it must be left to itself all of the time, or most of the time, or any other length of time. It only follows that *if* it is left to itself, it operates in a deterministic way. We can summarize God's intervention in nature on a few different levels. The first is God's role in the creation of the universe: God designed and set the universe in motion, as it were. This view is supported by the overall design in the universe, and the need to explain its origin from non-physical causes, as well as by the deterministic teleological effects that God has built into the universe, leading eventually to the coming into existence of conscious, rational observers who have free will and moral agency. There is also the level of God intervening in his creation through miracles, where miracles are understood as interruptions in the laws of physics. This level has two aspects to it: the question of biblical miracles (in salvation history), and the question of later miracles occurring in the natural world. Ernan McMullin has argued that the historical miracles of biblical times may be regarded as an exceptional form of God's activity in the world[4]; in any case, the biblical miracles can be judged on the overall evidential case for their occurrence, which is based on a combination of philosophical, theological, historical, archeological, and scientific arguments. Unfortunately, many are willing to make a judgment on these crucial matters *without* considering the evidential case. As for later and contemporary possibilities, it is true that of all the miracles that have been claimed in history, some are undoubtedly delusions, mistakes, or frauds, but the crucial point is that not all of them are, according to the theistic view. Of course, miraculous claims should be judged on a case-by-case basis, but the theist argues that there are many events for which there is no natural explanation. There are countless cases, in the area of medicine alone, of apparently miraculous events that suggest divine intervention in nature. Most theists hold that God does intervene in nature in this way from time to time (and that to rule out miracles *by definition* seems to make an unjustifiable appeal to naturalistic assumptions).

There is also the question of whether God intervenes in nature as a response to prayer; again this is entirely possible, and is accepted by most theists, not only as a possibility but also as a common occurrence. This intervention may also involve an interruption in physical laws, depending on the case (for example, in a miraculous cure from illness, but perhaps not in a case of a person praying for strength to overcome an addiction). Lastly, there is the level of God's possible intervention in the various causal chains in history. As we noted in our discussion in the previous chapter, this kind of intervention is possible, but whether it occurs or not is harder to make a decision about, partly because this type of intervention is simply hard to detect (and indeed God may have left no traces of his intervention).

Design in the universe

We now come to the question as to whether the presence of real chance within nature would undermine, as evolutionary naturalists usually claim, the conclusion that there is design operating overall in nature and throughout the universe? These naturalists are heavily invested in the idea of chance operating in nature; one reason for this is that they need the presence of chance in order to support their argument that there is no design in nature. They argue that there is no design involved, no teleology, in evolution, and then generalize from this to the view that there is no design operating anywhere in the universe. Dawkins, among others, has advanced this argument.[5] I think that Peacocke is correct in his general claim that the presence of chance in evolution would not undermine the existence of some design still being involved in some overall sense. The key question is whether it follows from the claim that if there are chance elements in nature, including in the process of evolution, then there would be no design operating in any other area of the universe, or in the universe overall? I have argued that it is hard to reconcile the nature of causation and chance, but suppose for the sake of argument that there *are* elements of chance in the universe, our question then is: is it possible to reconcile chance and design in the universe? Although, I have argued that there is no chance operating in the universe, it is not obvious that there is a logical difficulty in saying that God could arrange for chance and design to operate together. Many distinguished religious thinkers, in particular, are inclined toward this later view (as we have noted), including Peacocke, Polkinghorne, Ward, and Haught, among others.

The question of whether chance and design could coexist together is of pressing interest because the universe still shows evidence of clear design and intention in the complexity of organisms, and especially of *Homo sapiens*. One simply cannot ignore this incredible feature of the universe (as well as the evidence that is the basis of the anthropic, or fine-tuning, argument[6]), and it is not believable to argue that the emergence of such complex species is an accident of nature, and might just as easily not have happened. It follows from this that even if we accept that modern science reveals some elements of chance in nature, then the overall design that is nevertheless still apparent would leave us perhaps thinking along the lines suggested by Peacocke—seeking an argument that showed how chance and scientific laws work together to reach *the intended outcomes of the designer*. This is why, as I noted in Chapter 6, it is incredibly far-fetched to claim that a complex species such as man came about completely by an accident of nature (by "pure chance" as Monod put it). Even if one thinks that there were elements of chance involved, one still must deal with the fact that the end result looks like it was a designed, intended outcome. St. Thomas Aquinas expresses the general point well:

> We know from experience, however, that harmony and usefulness are found in nature either at all times or at least for the most part. This cannot be the result of mere chance; it must be because an end is intended. What lacks intellect or knowledge, however, cannot tend directly toward an end. It can do this only if someone else's knowledge has established an end for it, and directs it to that end. Consequently, since natural things have no knowledge, there must be some previously existing intelligence directing them to an end, like an archer who gives a definite motion to an arrow so that it will wing its way to a determined end. Now, the hit made by the arrow is said to be the work not of the arrow alone but also of the person who shot it. Similarly, philosophers call every work of nature the work of intelligence.[7]

The reply to Aquinas's reasoning that complex species, for instance, may *look* designed, but all came about by natural selection *operating by chance* both on the side of the organism and on the side of the environment, is not convincing in the absence of further argument, as I noted in earlier chapters. This is true even if we had the full evidence for natural selection because this evidence by itself would not show that the process operates *entirely* by chance. Keep in mind also that St. Thomas is arguing that it is not only complexity in nature that suggests design, but also ordinary objects and processes (including the physical laws); he agrees with Aristotle that objects in nature (such as an acorn) are directed toward an end (an oak tree). We discover this from our empirical study of nature; but objects (and laws) cannot direct themselves, so there must be an intelligence behind the universe. It is in this context that we should understand Newton's admission that his law of gravity did not really explain the *essence* of gravity, but only how objects behave when in a certain relation to each other. Teleology is evident everywhere in nature, according to this view, and the rational explanation for it is that there is a divine intelligence responsible for it.

In considering the question of design in general in the universe, there are three possible scenarios to think about. The first is one where we have a perfect world, full of optimum designs, and containing no imperfections (and so there is no problem of evil because there is no pain and suffering). The second possibility is that we have a universe dominated by complete chaos. This is a universe in which there is no order or regularity, where no living things can even come to exist, or if they can, they are unable to evolve beyond a rudimentary stage, where the process of evolution operating largely *by chance* never leads to any significant complexity, and especially not to conscious, rational, free beings. It is also a universe where there can be no stability in atoms or molecules, and where the formation of permanent galaxies and planets cannot occur. Of course, we would not be here if this kind of setup existed, but this does not alter the fact that this is one way the universe might have developed. The third scenario is our actual world, a world full of sophisticated, quite remarkable complexity that has led to conscious, rational, free observers, but also a world that has imperfections and challenges, yet not so much that they are fatal to the survival and development of complex species. The argument we are making here, and one that is echoed by many philosophers and theologians, is that the third scenario, while not leading to a perfect world, is nevertheless one whose structure and effects suggests design in bringing them about. It seems reasonable to say that the fine-tuning of the universe to produce life, and the two processes that come together simultaneously to produce the remarkably complex organisms that we find in nature, probably did not come about entirely by chance.

This was one of the main points Paley was making with his famous example of finding a watch in the wilderness (as noted in Chapter 2). Paley pointed out that after one had examined the watch carefully, figuring out how it worked and its purpose, one could not but conclude, because of its complexity, that it was designed. Moreover, this conclusion would not be disturbed if one began to notice imperfections (or design flaws) in the watch, because the evidence that the watch is designed is still very strong, indeed indisputable. I have no doubt that this would be Paley's response today to the theory of evolution. It is often thought that evolution refuted Paley's argument from design because he concentrated on the apparent design in organisms and their habitats in his argument for God's existence, both of which evolution can now explain. But Paley was also making a more general point, which he could adapt to include a response to the theory of evolution. Paley's general point was that the emergence through processes in nature of complex species, culminating in the arrival of conscious observers like us, with consciousness, rationality, and moral agency, looks designed. It is this complexity that makes us doubt that the process can be governed entirely by chance—because the evidence from nature (and today from evolution itself) is against it. Paley would point out that the blind alleys, setbacks, and imperfections in nature are analogous to those found in a watch, but they do not invalidate the inference to a designer. Even if one thinks that there is an element of chance involved, the end result is so complex that it suggests design. This is why it is crucial that if one wishes to claim that the complexity can be explained by natural selection operating *without any guidance*, one must show two things together: how the process operates *by chance*, and *how* it can produce complex organs and species. Promissory notes with regard to these crucial questions do not cut the mustard.

There are more general objections to the claim that evolution is an argument against the existence of design not just in nature, but in the universe as a whole. This brings us back to the notion of cosmic evolution. Some thinkers like Sagan, Dawkins, Coyne, and others are wont to talk as if evolution can explain *everything*, including the laws of physics! As Dallas Willard has noted, Dawkins and others are "in the grip of the romanticism of evolution as a sweeping ontological principle, bearing in itself the mystical vision of an ultimate *Urgrund* of chaos and nothingness of itself, giving birth to the physical universe, which is all very fine as an aesthetic approach to the cosmos, and vaguely comforting. But it has nothing at all to do with 'evidence of…a universe without design,' as [his] book suggests."[8] Alvin Plantinga has described evolution as an "idol of the tribe," the only game in town if you are naturalist.[9] It becomes a kind of mystical theory for secularists like Dawkins and Coyne because they need it to provide the answers to so many challenging problems for their view of reality. Dawkins's remark that before Darwin it was not possible to be an intellectually fulfilled atheist has been much quoted[10]; it seems to suggest that he thinks that evolution can explain all of the apparent design we see in the universe, including the laws of physics, even the origin of matter! We must remind ourselves again that the confusion between evolution and naturalism, between science and atheism, is all too evident in the work of these writers. A moment's reflection reveals that there are questions that evolution cannot help us with, and this includes our two biggest questions on the subject of God's possible existence: (1) how did the universe come to be, what is its ultimate cause?; and (2) how did the design of the universe come about, "design" here understood as the regularities present in the underlying laws of science (rather than the complexities in the nature of species). Evolution also runs into what seem to be insuperable difficulties in trying to explain key features of human life such as the origin of mind, and the nature of morality and free will, as we will see later. The question "what caused evolution?" should always give pause to those inclined to claim too much for the theory, to those who look at it as a kind of mystical explanation that somehow gives birth to the physical universe, and everything in it. Moreover, the coming into existence of any universe is an extraordinary happening, but the coming into existence of a deterministic universe that is also clearly teleological, and leads to complex, conscious, rational, self-aware, moral, agents, suggests a designer.

Intelligent Design theory

In recent years, the debate concerning evolution and religion has often been dominated by arguments about Intelligent Design theory (ID). While I have found the questions generated by the thinkers who advocate intelligent design theory quite fascinating, and agree that they have provoked a very interesting conversation with regard to evolution and related issues, such as the definition of science, it is important to distinguish my overall position in this book from intelligent design. I am not interested here in the political or social debate generated by intelligent design theorists concerning whether ID should be taught in public education, as a rival hypothesis to evolution, though the public discussion has often been dominated by this question. The fierce reaction to ID by many who seem to have joined forces against it, including leading evolutionary biologists, leftwing academics and atheists, is indicative of the political and moral divisions, indeed polarization, in our society. It says more about our disagreements in these larger areas than it does about the quality of arguments on either side of the scientific and philosophical debate. This is very unfortunate but is reflective of where our society is right now on these questions; it is also a reason for why much of the public opprobrium heaped on ID by its critics should be taken with a grain of salt. McMullin has noted the way criticisms of evolution in general "often evokes from its defenders a reaction reminiscent in its ferocity of the response to heresy in other days."[11] As we pointed out in Chapter 1, debates about evolution are often about a lot more than that; they often concern worldview questions, and political and moral issues, and one's views on these other matters tend to make an objective discussion very hard to attain, to put it charitably! I am also not interested so much in whether the main claims of ID are true or not. I am more interested in distinguishing my view from ID, especially in the light of my argument concerning determinism and chance.

I like to explain the main claims of ID theory in terms of three main points. These points can be found in the work of leading ID theorists, Michael Behe, William Dembski, and Phillip Johnson.[12] Generalizing somewhat, the first point these thinkers make is that natural selection cannot explain the complexity that one finds in cells at the molecular level. Behe has argued that the human cell resembles a small engine, so intricate are its parts and their interrelationships. Some extend this argument to include the claim that natural selection cannot account for very complex organs and biological systems, such as eyes and nervous systems. This first part of the argument is also at the same time a critique of the theory of evolution, especially the thesis of natural selection. ID theory becomes distinctive in its next two claims.

The second claim is that these "irreducible complexities," as Behe calls them, are best explained as having come about by the operation of an intelligent designer, rather than by a natural process. The third claim then is that this second claim is a scientific, not a philosophical or theological, claim, and so ID would be part of science, not philosophy or theology. This conclusion that ID is part of science is supported by the argument that biology shows evidence of design in an *empirical* way, so the claim that design is evident is based on a straightforward investigation of living cells through various experiments in molecular biology. In this way, the conclusion of an intelligent designer is arrived at in the way an archeologist might conclude that paintings in a cave were produced by an intelligent mind, or the way a detective might conclude that a fire was set deliberately (i.e., by an intelligence). Perhaps we should add a fourth point of clarification to note that ID is not the same position as, and indeed is not a form of, creationism because of two crucial points: first, proponents of the theory allow the *empirical evidence* to drive their arguments unlike creationists who usually begin from a religious perspective and allow the Bible to motivate their main points; second, supporters of ID do not officially identify the intelligent designer as God, but conclude only that there is an intelligent designer based on the evidence.

Sometimes critics will question the motives of ID theorists, claiming that they are religious in nature; yet, from a philosophical point of view, the motive behind an argument is not that significant, because logically one should look at the argument itself, rather than the motive of the person who presents the argument. The argument must stand or fall on its own merits, and be analyzed in terms of its logical structure and subject matter. This is true as much for an argument for intelligent design that one may think is motivated by religion as it is for an argument in favor of evolution that one may think is motivated by atheism. We do not discount the theory of evolution because some people who passionately advocate it, such as Dawkins or Coyne, may come across as fanatical, dogmatic atheists[13]; similarly we can't discount ID theory based on our perceptions of the motives of its advocates. To put this point in another way, we must distinguish carefully between someone who looks at various parts of nature in the way Behe has described so well, and who concludes that they might be designed, and someone who begins with the notion of a designer, and then looks at nature to (perhaps) over-interpret cases to confirm their initial hypothesis. Some may have the second approach, but as philosophers we must only consider the first approach in appraising ID, just as the fact that some theorists, like Dawkins and Coyne, *want* evolution to be true for various political and moral reasons should not dissuade us from making an objective appraisal of the evidence. Some might not trust the writings of Dawkins and Coyne on the question of the evidence for evolution because they often come across as dogmatic ideologues, but we can still discuss the theory of evolution objectively by reading widely and engaging in honest, open debate.

ID theorists have raised some difficult questions for natural selection, and their argument that the complexities in nature suggest an intelligent designer is interesting indeed. Their explanation of design in biology would seem to require an act of special creation by an intelligent designer who intervenes in nature at certain times to bring about the complexities, a view we discussed briefly in the previous chapter. This view differs from mine because although I raised difficulties for natural selection, especially with regard to the evidence for producing radically new species, my view is completely open to the possibility that natural selection (but not operating by chance) may be the way that complex species came about. My main claim is that the process operates deterministically, not that it might not have happened, though the latter issue concerns a debate about the evidence. My argument in this book is that whatever way the species came about, whatever the actual chains of causes and effects, it will be a deterministic process (the complex outcomes of which suggest a designer). So I would not agree with those ID theorists who argue that God intervened in a special way to bring about biological complexities.[14] Of course, it is possible that God could have brought about biological complexities this way. If God exists, he could intervene at any time in the processes of nature, but on my deterministic view it is not necessary for God to intervene because, given the initial starting conditions and the laws of science, the final outcomes, including the existence and complexity of species, had to occur. Another way to make this point is to say that, on my thesis of determinism, there would be no need for God to further intervene to bring about the outcome, because the outcome *must* come about given the initial conditions. Intelligent Design theorists do not hold to a deterministic view of reality, and they also appear to hold that there is some element of chance in nature. But, despite this element of chance, they believe that direct intervention by an intelligent designer is needed to bring about the *specific* complexities in organisms, and that no natural explanation is likely for these.

Many have argued that this leaves ID particularly vulnerable to a "god of the gaps" type of objection. The "gap" is our present inability to give a scientific account for the biological complexities, but the danger is that eventually we will discover an explanation, and so ID will be left embarrassed. Behe is not worried by this criticism, and has argued that the more we come to discover in biology at the molecular level, the worse things are getting for natural selection. This criticism is also related to a second common objection to ID, that it is not real science, that its claim to be a scientific conclusion, or even a scientific research program, is bogus. This is because the discipline of science should be confined to physical causes and explanations (a kind of definition of science[15]); so, any explanation that involves an appeal outside of the physical order should not be counted as science. Why is this? It is possible to perhaps identify four possible reasons. The first is to say that this is how science has been traditionally understood. This is perhaps true, but it is not really a strong philosophical objection, because one can always challenge traditions. A second reason is based on an assumption of naturalism, where one holds that the *only* kind of explanation one will allow is a physical explanation for any phenomenon. This objection is obviously question-begging. (But it is worth noting that some ID theorists, especially Phillip Johnson, argue that the official theory of evolution, as it is presented today by leading theorists and in most biology textbooks, assumes philosophical naturalism, especially in the denial of teleology. This is despite the occasional protestations of some of its official spokespersons to the contrary, and despite the goodwill of many scientists who reject and regret this tendency.)

A third possibility is to argue that there are criteria for judging a good scientific theory, such as falsifiability, testability, coherence with other theories, and ability to make predictions, and that ID does not satisfy these criteria. This is a better criticism than the other two but one problem with it, as many have pointed out, is that there are many theories within science that don't satisfy these criteria either (indeed some claim that evolution itself is one such theory since it can explain, but cannot predict,[16] and also, as discussed in Chapter 3, it is very hard to falsify it—it may fail Popper's criterion of falsifiability). In addition, when subjected to rigorous philosophical examination, the criteria themselves are far from clear-cut. The fourth reason is one of the main reasons offered today against ID. This is the view that there is a consensus within science that ID is not science, and that evolution is true. The fact that there is a consensus is probably correct, especially with regard to the truth of evolution. But how much weight should be given to this type of objection, to the claim that because many scientists think that ID is not science, then it is not science?

The answer partly depends on how one understands the nature of the evidence for evolution. It is often claimed, for example, as another objection to ID, that natural selection *can* explain the complexities involved in nature, and so the inference to a designer is not justified. This has been the approach of critics like Kenneth Miller.[17] The claim that evolution can explain the complexities can be understood in two ways. It might mean that we actually have explanations now for how complex organs and mechanisms in nature came about. If this is what it means, it is not true, and this area is often a source of much exaggeration in biology textbooks, and involves glossing over gaps and difficulties in the argument.[18] A second way to read the claim is to say that although we can't explain the complexities now, we will be able to explain them soon. This latter claim can itself be understood in two ways. It might mean that we know so much about the process right now that we are very close to a breakthrough (in the way one might claim that we know so much about how engines work that we will very soon be capable of constructing them to be much more efficient in terms of fuel consumption), or it might be a more general "god of the gaps" claim in its own way—meaning that we have little idea of how it works, but we have faith that we will find a scientific explanation eventually. This latter interpretation is closer to what biologists mean when they say that we will eventually explain how complexities came about in nature. The first of these readings would certainly give any reasonable person pause, but it is easier to reject the second one as an unwarranted assumption, an exaggeration, or even as wishful thinking. It is for reasons such as these that ID has gained a foothold in the discussion, and indeed has forced evolutionary biologists to try to meet its objections by focusing on the specifics of natural selection, and confronting the gaps in their previous accounts.

Even though I would reject the notion of God intervening directly to bring about the complexities because it is not required by my deterministic view of the workings of nature, and even though I would also keep to the traditional understanding of science, I welcome the very interesting, probing questions that ID has introduced into the discussion about evolution, science, religion, secularism, and philosophy. Leaving all of the political and ideological baggage aside, ID has generated a fascinating and productive discussion. I welcome also the way it has forced evolutionary naturalists in particular to be more careful about what they are claiming, and chastened them to be more honest about the evidence they have for these claims. And let us not forget that while the public controversy about ID has often been about whether or not it should be described as science (and so discussed in biology classes), this is ultimately a secondary matter because it is really only a debate about how ID is to be classified, whereas the real question must be about whether or not it is true.

Determinism and the question of evil

We must now attempt to elaborate the implications of my thesis of determinism concerning the universe and the processes in nature for the imperfections in creation, for those elements that do not appear designed or intended, in short, for the problem of evil. Arguments against the existence of a designer based on the problem of evil often appeal to the theory of evolution, and specifically to the pain, hardship, and waste in nature that evolution reveals. One form of the argument is that the waste and inefficiency in the process of evolution may be a way to argue that evolution operates without design, even if we were to grant our argument that there is no chance involved in the process. The idea is that the waste and suffering in nature is a bad thing, and that no designer would have planned it out this way. Part of the argument is that evolution should make us reappraise the problem of evil argument against God's existence, that the theory should lead us to conclude that the problem of evil becomes *worse* after evolution than it was before evolution, and that this is further evidence against the existence of God. Before we look at how evolutionary theory is supposed to have significant implications for the argument that God does not exist based on the problem of evil, we need a brief reminder of what the problem of evil is.

The problem of evil for God's existence is based around a simple, but very challenging, question: why does an all-powerful, all-good, and all-knowing God permit evil to exist? Evil is understood to refer to both natural and moral evil. Natural evil is the evil that occurs in nature that is outside of man's control, such as earthquakes, floods, famine, and disease. Moral evil refers to the category of evil that is the result of free human actions, such as murder, robbery, and rape. The probabilistic or evidential form of the problem of evil argument (which is the form we will consider here) says that natural and moral evil must count as evidence *against* the existence of an all-good and all-powerful God, and so make it *probable*—more probable than not—that God does not exist. This is because it seems reasonable to think that if God's nature is perfectly good, then he *would* want to prevent evil (if he could). But God is also supposed to have the power to prevent evil, because he is omnipotent. And yet we have lots of evil in the world; so, according to this argument, there is likely no God. The argument does not claim to prove that there is no God, but only that it is more probable than not that there is no God.[19]

In what way does the theory of evolution appear to make this argument stronger or worse than it already is for the theist? (The theist does not deny that it is a serious challenge, but does hold that there are reasonable replies to it.) Let us look at three ways in which some thinkers have argued that evolution throws new light on the nature of evil (also keeping in mind our thesis of determinism). We will focus on natural evil in our discussion, since it is more relevant to the question of evil occurring in nature itself. The first way is to suggest that the process of evolution reveals that there is a great deal of waste in nature, that the existence of natural evil is on a much larger scale than we had realized. Before evolution, we understood natural evil primarily in terms of how it affected us as human beings. So we would regard an earthquake or a flood that kills many people as evil because it causes so much human suffering. But after evolution, we are now forced to consider other kinds of evil: the waste, pain, and suffering that occur in the animal kingdom, and the all-pervasive waste that occurs in the natural world in general. One area in which this waste is obvious is in the large number of extinctions that have occurred since life began three and a half billion years ago. Over 90 percent of species have become extinct, according to some estimates, and all present species (including us), it is claimed, will very likely become extinct.

Another way to think about the waste in nature is to focus on the general predatory nature of life both at the plant and animal levels. Evolution shows that there is a struggle for survival going on in nature, often in situations of limited resources, and that many species and individuals live off of other species and individuals. Darwin was struck by the predatory nature of animal existence in developing his theory, and he referred to the example of the ichneumon wasp, a parasite that feeds off the bodies of caterpillars in its attempt to survive, as a case that made him question the idea that nature has a purpose, and is created by a benevolent deity.[20] A study of species survival reveals a constant predatory and parasitical battle with other species and individuals leading to hardship, suffering, and eventual extinction. This seems to be a defining part of the nature of life; it is illustrated clearly by the process of evolution. In addition, we did not know this before the theory; we knew only that life was often difficult for us (and we did not usually consider what happened in the rest of nature, so it is claimed). One might say that before evolution we looked at natural evil only in terms of how it affected us, but after evolution this way of looking at it is no longer accurate or adequate. In addition, if the thesis of determinism is true, it would mean that God intended in some way for all of this natural evil to occur, up to at least the appearance of us. If God set up the initial ingredients of the universe, together with the laws of science by which matter and energy behave, then all outcomes are predetermined (as I argued in previous chapters). And so would it not follow from this, some will no doubt argue, that the evil that occurs, especially in the animal kingdom, but also a lot of evil that affects human beings, was intended from the beginning, since it has to come about? If nothing random occurs in the universe, the existence of evil would not be random either. One might then extrapolate from this that an all-powerful, all-good God does not exist, and that there is no design in the universe.

A second line of argument that the theory of evolution raises that is significant for religious belief is the matter of imperfect designs in nature.

It is often claimed that because the process of natural selection is random, haphazard, and affected by so many variables, which are themselves random and haphazard, that we end up with many adaptations in nature, in various species and individuals, that are less than optimum for that species or individual to survive (or to live without hardship). Natural selection does not necessarily produce perfect adaptations, but only adaptations that "work" — adaptations that enable the various functions in an organism to operate reasonably well, which aid not only survival but are also determinative of whether or not the organism has to endure hardship. One of the examples given to support this view is the human eye, the subject of much discussion in the literature on evolution (and also in the debate between evolution and Intelligent Design theory).[21] We have noted the argument that a structure like the human eye is so intricate and so complex, and requires so many sophisticated parts working together at the same time, that it must be designed, and that the thesis that it came about through natural selection operating *randomly* in nature is far-fetched. This argument looks very plausible to many people. One of the arguments some evolutionary naturalists have advanced against this view is that the eye has imperfections that are hard to explain if one holds that it has been designed. In addition, secularists will argue that the imperfections are further evidence of the presence of natural evil in nature.

An example of a significant imperfection in the human eye, it is claimed, is the blind spot. The blind spot is caused by the optic nerve running through a hole in the eye, according to Burton Guttman.[22] This development occurred during the process of natural selection when the blind spot was forming. Because the process was random and inexact, and subject to the vagaries of the existing (prior) structure of the organism, as well as the haphazard nature of its environment, eyes began to form in significantly random ways. They developed whatever way they could, which would inevitably be less than perfectly. And some organisms in which an eye developed or was developing survived and some did not. But those that did survive did so because the eye, despite its imperfections, was still partially operative and so gave them an advantage in the struggle for existence, and aided their survival. It is not hard to see how the existence of sight, even if it were not perfect sight, would give an organism a clear advantage over one that did not have sight, and so give it an edge in its struggle for survival. But if the eye was designed, Guttman contends, there would be no blind spot; the blind spot, he speculates, is due to the haphazard way the eye and the optic nerve formed. Of course, the blind spot is not a huge blemish on sight, since eyes work well in organisms even with the blind spot, but it is still an imperfection caused by nature, and in some cases it can lead to the death of the organism (e.g., in leading to accidents, or in being unable to see danger). Other examples of imperfections often referred to include the fact that in our breathing apparatus the trachea is on the wrong side of the esophagus, and the panda's "thumb," which is actually a wrist bone adapted (so speculates Stephen J. Gould) due to the panda's diet (which is relative to the panda's environment).[23]

Although Guttman's form of the argument relies on the operation of chance elements in nature, one can adapt the argument to take account of my position in earlier chapters where I tried to show that changes in both mutations and the environment do not come about by chance. One could argue that even if the operation of nature is deterministic, including the process of evolution, it would then be preordained that some creatures would have imperfections. The process also results in scores of species not being able to survive, or surviving for only a short time in hardship, or having to endure much suffering, and then dying out, and this by itself—even if the whole process is determined—would still count against design. One might hold that if the process was determined and it arrived finally at a perfect eye, or species, or organism, that would be one thing. But if there are many imperfect features in many organisms along the way, with many of the imperfections leading to pain, suffering and death, and there are loads of half-baked, improperly developed species, and some that feed off others, etc., doesn't this undermine our confidence that the process is designed, that it is the intended outcome of a plan? An atheist could grant determinism, but then argue that the starting point must have been chaotic in some way; this is why the development of life is so disjointed, inefficient, haphazard, imperfect, etc. The argument is that since there are imperfections in nature, nature is not the product of design. If God had designed nature, either directly (he designed the eye directly at the appropriate point in nature, the view of the ID theorists, the position known as "special creation") or indirectly (he set a process in motion with in-built mechanisms that would produce specifically intended results, including organisms with eyes), then we would not have imperfections of this type, imperfections that can cause suffering and hardship, and problems with survival, and so forth. So this is usually also a general argument for atheism.

The third argument is to reject the thesis of determinism about nature that I have been advocating, and to appeal to the operation of chance throughout the process of evolution in a more general sense. It is often claimed by secularists who co-opt the theory of evolution to develop an argument based on evil that much of the evil that occurs in nature is due to chance. This means that not only is the *amount* of natural evil hard to explain, but its very existence is hard to accept once we realize that it originates by chance. Evolution adds this extra point to our understanding of the problem of evil.

In short, one might interpret the *randomness* in nature as itself a form of evil, and not only as showing that most evil events originate in nature because of chance. But the *random* nature of the events can be seen as an underlying layer of evil in nature. Evolution shows randomness and chance occurrences on a large scale in nature, and this itself is a form of evil, one can argue, because it is not what one would expect to find if an all-good, all-powerful God exists. As Dawkins has noted, we like to think that things are working out for the best, or at least that they are moving toward some end or purpose, but when we look at mutations we see that they occur without any end in mind for the species, without regard to how they could benefit the species, as he puts it.[24]

Gould also supports this reading of the evidence from evolution, though, unlike Dawkins and others, he does not press it in the service of an objection to the existence of God based on the problem of evil. But, as we have seen in our earlier discussions of Gould's view, he thinks a large element of chance operates in nature, and this colors his interpretation of the process of evolution. He refers to several examples used by Darwin in an exchange with American scientist, Asa Gray, on the topic of the hardship present in the animal world, including the examples of a man killed by lightning on a mountain top, of a gnat being killed by a swallow, and so on. Gould's and Darwin's view is that there is no overall meaning in these events; they simply happen randomly, though, interestingly, neither uses them to argue that God does not exist, only to argue that God does not intervene in nature, or imbue every action and event in nature with an overall, deeper rationale or meaning (though Gould sometimes confuses the notion of the actions having a deeper meaning with the separate point that God intended for them to occur).[25]

By way of response to some of these challenging problems, we must point out that, first, if our extensive discussion of chance, randomness, and determinism in our previous chapters is correct, then we can dismiss the popular but erroneous claim that mutations and other changes occur by chance in the process of evolution. Applying this conclusion to the problem of evil allows the theist to counter the third argument discussed above. The theist can reply to the point that there is no underlying rationale for the evil in nature by arguing that this only seems to make sense if it is true that evolution operates by chance, but it is not true if all of nature is determined by the existence and nature of the initial conditions (and the laws of science). This is because the determinism that we have been describing suggests that there may be an underlying purpose in the universe after all—a type of order or design, if you will—a goal or set of goals that it is moving toward, and so what happens in nature, including evil, may all be an intended part of this purpose.

We may not be able to discern the purpose completely, but the discussion just brings us back to the question of who started it all off, and why. And so therefore it is a very interesting point indeed that the evolutionary process gave rise to complex, conscious, intelligent observers who have some control over the process. If there is no chance involved in evolution, Gould would be wrong in saying that because the existence of every species is due to chance, no particular species had to come into existence; and Dawkins would be wrong that the final structure an organism ends up with is due to chance (not even to mention the exceptional qualities of *Homo sapiens*, reason, logic, free will, and moral agency, that create very difficult problems for evolutionary explanations, and that we will return to below). The naturalist may press the point that we are then saying that all the evil in nature is part of God's purpose. But this objection misses the mark because this is what the theist held *before* the theory of evolution, and is why we are able to generate the problem of evil in the first place.[26] It simply leaves us back where we started with regard to the problem of evil, but this time *after* evolution, so evolution makes no substantial difference to the nature of the problem. This problem of what God's purpose may be in allowing evil is difficult to solve, and our aim is not to solve it here, but the argument I am trying to refute is not that there is a problem of evil, but that neither determinism about nature, nor the theory of evolution, makes the problem of evil worse.

Some modern theologians think that they can blame the existence of evil on the role of chance in the process of creation (and so this would be an advantage of their view over my deterministic view). But it does not follow that the thesis of determinism makes the problem of evil any worse for my view than it is for a view like Peacocke's, who holds that there are elements of both chance and determinism involved in nature. This is because the latter view still has to address the question of why evil events occur along the way toward the intended final outcomes. Even if one attempts to argue that the *chance* elements are responsible for evil happenings (as Peacocke, Keith Ward, and Francisco Ayala do[27]), this argument is far from convincing unless one can show in specific cases how chance elements in specific causal processes led to the occurrence of evil events; and, in any case, wouldn't we still have to explain why God introduced chance elements into a process that can lead to evil events along the way? Would God still not be responsible for the overall process, if he set it up at the beginning? Peacocke and Ayala seem to believe that the presence of chance elements would somehow absolve God of responsibility for the existence of evil, as far as I can see, but it is not clear how to understand this if God designed the process that allows for the chance elements that lead to evil, yet directs the final outcomes to a significant extent.

This brings us then to the first argument mentioned at the beginning of this section, the objection that evolution shows that there is a lot of waste in nature, which makes the fact of evil worse than we imagined. In response to this objection, we must acknowledge that there is a lot of evil, and that it is indeed a problem for the claim that an all-good and all-powerful God exists. Nobody denies that. Yet it is not clear that the theory of evolution really adds anything to the problem. In particular, it is not clear that it does show us that there is more evil in nature than we originally thought, or at least not significantly more that would lead us to think that the problem of evil was worse than we imagined. For before evolution we already knew that there was significant waste in nature, as well as much suffering and hardship. We also knew about extinctions.

We must also not forget that many of the specific claims about how specific changes in nature came about through the evolutionary process are based on guesswork, such as the number of species that became extinct. Without specific evidence in specific cases, we are dealing largely with impressions and "just so" stories about how much waste there is, about how the process is inefficient or how features are badly designed, some of it driven by an ideological agenda, and so we should be wary of accepting these speculations as the basis of a serious new insight on the problem of evil. It is also not clear how we should define the notion of "waste" in nature. When we say that many species become extinct, we are inclined to see this as an example of waste. But how is waste to be understood? It is difficult to place a value on a species, especially if we attempt to define it in terms of long-term health or suitability for its habitat, or its lifespan (compared to what: other species? age of the planet, etc.?), or its usefulness (again, in terms of what: other species? us? its competitors? nature in general?). These are hard questions to answer without appealing to some general theory of the value of existence or meaning of life, and then we may have a circular argument on our hands. We may run into the real danger of using a general theory about the meaning of life to arrive at a general theory of the meaning of life. We could appeal to an alternative, more medieval perspective that "all of nature is good" to argue that the existence of all of the species, however short their lives and whatever their imperfections, is nevertheless an invaluable part of reality (this argument should particularly resonate with environmentalists who appeal to a similar argument today to defend protection of the environment even if it costs hardship to humans). In short, any claim about whether something is a "waste" in nature, especially if it means that some lives have little value (for whatever reason) already assumes a general theory of the meaning of life, and so would seem to preclude interpreting certain alleged cases of waste in nature as supporting one general theory over another. To put this point another way, it is very controversial to claim that scientific evidence alone allows us to conclude that the existence, nature, and lifespan of any species, or member of a species, is a "waste" in any clear sense of that term. Furthermore, the very fact that an organism exists means that it should not be regarded as a waste, and this would be especially true from God's point of view.

Similar points can be made in response to the second argument considered above concerning certain imperfections in nature, such as the blind spot of the eye, accepting for the moment, for the sake of argument, that the blind spot originated as a byproduct of inefficient, random evolutionary processes (though I suggest that we have read enough evolutionary "just so" explanations offered by evolutionary theorists to know, not only that they are very speculative, but that are *not* the way things happened). But the fact that the eye apparatus is *not* perfect seems to be on a par with saying that the operation of the heart is not perfect because the arteries can become clogged, or that hip joints are not perfect because they wear out, or that, more generally, the body is not perfect because it is subject to disease. These are all problematic features of nature that force us to raise the problem of evil in the first place. And I have acknowledged that no theist wishes to deny this, or to downplay the fact that natural evil gives rise to a legitimate question about God's existence. But my point is that the evidence from evolution does not make this problem any worse than it was before the theory of evolution. For, like the blind spot and other blemishes and imperfections in nature, we already knew that nature was blemished with flaws, limitations, inefficiencies, and other problems that all make up the problem of natural evil. Evolution reveals a few more of these flaws, but this seems to be akin to the discovery of new diseases, or to the discovery of the underlying cause of an already known disease perhaps, but it does not alter fundamentally the nature of the problem we are dealing with, or show an extra dimension to it that makes it worse than it was before. We can see the imperfections in nature as part of the general problem of natural evil, not as a new form of it, and not as altering our understanding of it in a fundamental way.

A naturalist who had co-opted evolution might reply that one new thing we know from evolution is why the blind spot, for instance, formed, and we know that it formed in a haphazard way. But we have already replied to this point. First, we don't really know how it formed; our account is based on speculation, which may or may not be (and almost certainly is not) accurate. Second, even if we had agreement on how it formed, it would still seem akin to saying that we know why people get heart attacks—there is a causal chain that leads to the heart attack as an end result, and we can ask why these kinds of chains leading to evil events occur at all, but this is the original question that motivates the problem of evil, and not a new form of the problem. And third, if my argument about there existing a certain determinism in nature is correct, then the formation of the blind spot would be an intended outcome (or side outcome) of the process of the evolution of human sight, and we must acknowledge that it does force us to ask the question why. But does it make the question any more difficult (difficult in kind and not in degree) than it already is? I don't see that it does. It simply returns us to the problem of evil itself, which we will not discuss here. That is a task for another time, though it would involve two main points: that God has a reason for natural evil, and that the free will defense can be invoked to explain moral evil (see Endnote 26). In any case, the key point for our discussion is that the problem of evil is not a problem that is made significantly worse by either the thesis that nature acts in a deterministic way, or by the evidence from the process of evolution.

Evolution, naturalism, and morality

The objection that motivates the problem of evil argument against the existence of God can be thought of as a moral objection, even though it is often presented as only a logical objection. That is to say, it is usually presented as saying that if the evidence for natural evil is strong, then it would not be reasonable from a logical point of view to say that God exists. But occasionally its supporters are thinking more of a moral objection along the lines that the evil in nature is *morally wrong*, and that God, if he existed, would not allow it. Since it exists, he probably doesn't exist. What is interesting about this form of the argument is that *it assumes that there is an objective moral order in nature*, and this assumption raises further problems for those who would reject theism and embrace atheistic naturalism. This is because this kind of argument, and any argument for atheism based on the problem of evil, assumes that evil of this sort is wrong, and should not happen if there is a designer. Of course, if there is no designer, and there is no overall meaning or purpose to the universe or to life, then this raises a fundamental question of how any kind of objective moral order can be justified. How is it possible to judge a moral action or behavior as objectively right or wrong? For those who co-opt evolution to support their atheistic naturalism, the question is how it is possible to justify morality from the perspective of evolutionary naturalism? In short, if God is not the author of the moral order, then where does it come from? Can we justify the existence of an *objective* moral order if it originated out of an *unguided* evolutionary process; can we justify objective moral values coming from a process that is not itself moral or immoral? It is important to emphasize that this is a question about the justification of the moral order. We are not asking if we currently believe in an objective moral order, or act on the basis of objective morality; we are asking how this objective moral order is justified, given its origins.

The first issue to consider is the problem of moral relativism. Moral relativism is the theory that there are no objective moral values, nothing that is objectively right or wrong in itself, irrespective of human attitudes, opinions, customs, or cultural traditions. Proponents of this view hold that moral values are not objective in themselves but are relative to some other feature of experience or nature, and this feature gives them their identity (just as the temperature in a building is relative to the outside temperature). Moral relativism comes in two forms: extreme relativism and cultural relativism. The position of extreme relativism (which is also sometimes called subjectivism) holds that moral values are relative to the individual person, and are based on one's subjective opinion, tastes, individual interests, and so forth. The cultural relativist holds that moral values are relative to one's culture (which is often, though not always, identified with the dominant moral views in one's country). A very significant consequence of moral relativism is that one is not entitled to criticize the moral values of another person (or culture), without falling into contradiction. This is often expressed more popularly in phrases such as "I have my values and you have your values," "who are we to judge the moral values of another culture?"; "who's to say what is moral or immoral?," and so forth.

Moral relativism has sounded very attractive to people as an alternative to moral objectivism, but it comes with two very large problems, both of which are considered by many to be fatal to it as a position about morality. They are really two sides of the same problem. The first side I call the practical problem. This is the fact that it is virtually impossible for us to live in society with others without making objective moral judgments about the beliefs, moral positions, prejudices, political beliefs, etc. of those we disagree with. (If you disagree with this, try living for a few days without making a moral judgment!) But to make a moral judgment of course is to contradict your relativism because it means that you are not practicing what you preach: that moral values are relative, not objective. The other side of this problem I call the logical problem of relativism. We can illustrate this problem by considering this statement: "It is {morally wrong} to criticize the moral views of others because there are no objective moral values, and who is to say what is right and wrong." The problem inherent in this view is that the "morally wrong" in the first part of the statement is an objective moral judgment! And so it leads to a contradiction because one is saying on one hand that there are no objective moral judgments (and so we should not criticize others) and yet, on the other hand, one is making an *objective moral judgment* oneself!

The reason moral relativism is related to the discussion concerning evolution and religion in an important sense is because we must raise the question as to whether any *naturalistic* account of morality must inevitably end up in moral relativism, including evolutionary accounts. It is often noted that it is no surprise that we live in what we might call "the age of relativism." This means that at the philosophical level we have lost confidence in the foundations of knowledge, not just in ethics, but in a variety of areas, including religion, politics, law, education, even science. A certain skepticism is at the heart of many academic disciplines today. Many philosophers doubt whether human knowledge can be given an ultimate foundation; this is why the dominant view in metaphysics and epistemology is antirealism, even in the philosophy of science. One also sees the influence of this position in the area of religion, with the view now widespread that no religion can claim to have the truth, that one religion is just as good as another. This view can also be extended to political theories, and to other areas of life, such as claiming that there are no objective standards in education, that there is no independent way to judge the university canon, that there is no one correct way to read a work of literature, and so forth. We must be careful to acknowledge that although it is accurate to describe these views as reflective of the current *zeitgeist*, by no means does everyone subscribe to these relativistic and antirealist views; we must also be very careful to distinguish between relativism and skepticism as theoretical claims, and the unwillingness to carry them through to their logical consequences (i.e., to follow them consistently).

The important question we need to consider is how this age of relativism came about. It was not just because of the critique of religion; it was because when one turns to naturalism to answer many of the questions to which religion used to provide answers, one discovers that it is unable to answer them. This is especially true in the area of morality. And so from the point of view of evolution specifically, which is our main concern here, the key question is: if one accepts an atheistic view of reality, and utilizes a naturalistic interpretation of evolution as one's main argument, can one avoid moral relativism? Or to put the question the other way round: is it possible to *justify* an objective moral order on a completely naturalistic evolutionary account of its origins? (A reminder that the theory of evolution is not naturalistic in itself, but is interpreted this way by those who would then have to explain the origins of objective morality.)

Let us try to identify the problem in more detail. If one holds that there is no design involved in the universe, or in the process of evolution, that all of the species and their features came into existence by chance (and perhaps in addition that evolution is not progressive), then how can one still insist that the moral order that has evolved over millions of years and that has been inherited by *Homo sapiens* is an *objective* moral order? There are a number of distinct concerns here. First, there is the question as to how an evolutionary process operating largely by chance could produce a moral order that would be in any way binding on everyone. What would make it *binding* if there is no orderer, no designer, no plan laid down in advance, but it is simply the outcome of a chance process? We can agree that evolution operating by chance could produce a moral order, just like it could produce a living being with legs, or a being with a heart and lungs, or a conscious life form, but one cannot argue in these latter cases that the organism has to have these features, only that it just so happens to have them. Would it not be the same with the moral order? So it seems that we could question why we have to adhere to it, since it came about by chance. Analogously, one could argue (and many do) that although we evolved with two distinct sexes and a view of sexuality as being primarily for reproduction, we don't have to adhere to this understanding bestowed upon us by evolution, and can change nature around to any sexual arrangements we like.

A second concern is more directly related to relativism. This is the view that relativism follows from an evolutionary account of morality in two senses: the values we end up with from the process are *relative* to the way history unfolded (in Gould's sense); if it had gone differently we would have a different set of moral values, and so there is no necessity about morality; the second sense is that we can therefore abandon the objective moral order we have now, and see it for what it is: a mistaken belief about the nature of morality. We attribute necessity to a process that is in fact governed by chance and contingency; so we regard murder, rape, and robbery as objectively immoral, but how can they be if the evolutionary process could just as easily have produced a different set of moral values? It does not seem to be a good reply to this argument to say that whatever objective moral order evolves, human beings must adhere to it. For how can one support such a view, since the resources one would need to do so are undercut by the claim that the objective moral order evolved by chance? Some naturalists, like Dawkins, have willfully misinterpreted this problem. They will reply to these points by saying that if one becomes an atheist, and subscribes to an evolutionary account of morality, does this mean that one would go out and start doing bad things? Of course, it does not mean this, but this objection misses the point. The point is that on an evolutionary naturalist view *one loses the foundation* for the moral order that one is following. This does not mean that one would not continue to follow this moral order, especially if one's character has been formed in accordance with it, in the way brilliantly described by Aristotle and St. Thomas Aquinas. But perhaps we should also note that the commitment of some people to their objective moral values and to doing the right thing probably would weaken if they came to believe that there is no objective basis to the moral order. And it must be said that this is how many people would interpret the moral direction of the latter half of the twentieth century.

The theist argues that the moral order is grounded in the existence of God, who is the author of the moral order. The moral order issues from God's nature, and so holds by necessity. This philosophical view also complements the biblical view that man is made in God's image, and one recalls St. Thomas Aquinas's view that the moral law is written on the human heart. This perspective has become the basis for the moral argument for God's existence, one form of which is: "We know that there is an objective moral order and objective moral values, so we can infer from this that there is an orderer, or author of the moral order." This argument has been advanced by Immanuel Kant, John Henry Newman, and C.S. Lewis, among others.[28] However, crucially, it is not available to the evolutionary naturalist, since naturalism implies moral relativism. In short, if naturalism is true, then moral relativism must be true; but moral relativism is not true, therefore, naturalism is false. I regard this problem about morality as a fatal one for any view that tries to explain objective morality totally in terms of a naturalistic evolutionary process. A purely evolutionary naturalist account undermines the objectivity of ethics just as it undermines the objectivity of everything: including the existence of natural kinds in nature (rather than accidental species)[29]; the unique nature of human intelligence; human nature (traits that are not reducible to, or fully explicable in terms of, purely biological processes); human essence (there is a unique set of characteristics that make up a human being); human uniqueness compared to other species; the distinction between the natural and the unnatural; and so on.

Understandably, some evolutionary naturalists shrink from these conclusions, or try to avoid them, but a few acknowledge them honestly, including Michael Ruse and Edward O. Wilson. Wilson has pioneered the field of sociobiology, the study of the biological basis of social behavior, using evolutionary principles as a guide, and has suggested that human beings are shaped more by their genetic inheritance, and their environment, than by anything else, and that there is no such thing as free will. Morality, therefore, is also the product of culture, which in turn is mostly the product of genetics; in addition, moral behavior must be explained in terms of its evolutionary advantages.[30] Michael Ruse has acknowledged that "Darwinian theory shows that, in fact, morality is a function of (subjective) feelings; but it shows also that we have (and must have) an illusion of objectivity." He suggests that while morality is "a collective illusion foisted upon us by our genes," it can perhaps be "explained" as an evolutionary adaptation![31] Dawkins has admitted that objective morality is a kind of illusion.[32] Although we all try to follow what we believe to be an objective moral order, Dawkins agrees that it has no absolute basis from an evolutionary point of view (and it is a form of enlightenment to realize this); nevertheless, he thinks that he will be unable to divest himself any time soon of the view that it *has* an absolute basis, and so he thinks it follows from this that we cannot help regarding morality as objective (at least Dawkins's version of it)![33] (Dawkins's book, *The God Delusion*, is full of moralizing to the point of irritation.)

There is another dimension to the challenge of the objective moral order for secularist views of reality. Even if we grant to evolutionary naturalists for the sake of argument that on their view one can still support the existence of an objective moral order, there are still many values that are part of that order that seem to run *against* an evolutionary explanation. This is particularly true of values such as altruism, love of neighbor, self-sacrifice, Christian love, and the traditional view of the human person, which puts the emphasis on the social nature of persons and on the community, rather than on one's own individual interests. These values appear to run contrary to evolutionary theory because they appear to go against natural selection. The main point is that evolution seems on the whole to reward selfish behavior—this is how the whole theory works.

Let us not forget that evolution is a "survival of the fittest" theory; a species that has a selective advantage over others (relative to the environment) is more likely to survive, and to thrive. If one attempts to explain morality along evolutionary lines, one would have to argue that morally bad behaviors must have had a selective advantage for some of our ancestors in their particular context. Isn't this how the theory of evolution explains *all* of a living thing's traits and behaviors? In addition, on this view, more selfish behavior on the part of an individual would be more likely to lead to survival and success than more altruistic behavior; in general, looking after oneself would be more beneficial than looking after one's neighbor. Of course, it is different with lower level animals in that they do not have a moral sense (a point of great significance, that we will turn to in a moment), and yet there are countless cases of altruistic behavior in the animal kingdom (e.g., some species of birds help other birds to feed their young). But when we get to humans, from a purely evolutionary naturalistic view of evolution, one can argue that our moral values are just another tool in our set of survival skills (indeed Ruse and Wilson are arguing as much), just like our limbs or our intelligence. Ruse has noted that we should no more get rid of objective morality than we would our eyes, but it does not follow from this that we could not abandon objective moral values (i.e., basically act like moral relativists), when it would seem to advance our own development and survival; nor does it even seem to follow that we could not get rid of objective morality altogether. Indeed, thinkers like Dennett and Dawkins have stridently tried to explain *religious* beliefs this way, arguing that such beliefs evolved and survived because they must have given humans some adaptive advantage in the struggle for survival.[34] So why not offer the same explanation for moral beliefs, and draw the same conclusion—that they are false, and we should get rid of them?! On this view of morality, one could argue that we should not be monogamous and faithful to our spouses because we can enhance our reproductive fitness more if we are promiscuous, and so forth. And that we should not help people with whom we are in competition (say at work) because this only harms ourselves; one could argue that our appearance is more important than our character because it helps us attract a mate, receive favorable treatment, is socially advantageous, etc.; that stealing is better than working (if one can get away with it, of course, as the Greek sophist, Antiphon, is alleged to have argued!). Indeed, such views motivated the program of eugenics that some thinkers developed based on Darwin's views.[35]

This general approach can be used to support a kind of nihilism about life and meaning, as several naturalists have acknowledged. Evolutionary naturalists are aware of this problem and have tried to respond to it in very speculative ways by arguing, for instance, that perhaps altruistic behavior helps us over the long run, even if not in the short run, or by arguing that cooperation at the group level can have advantages in terms of survival. The problem with these sorts of explanations is that there are thousands of instances in ordinary human life where being selfish would clearly benefit an individual in an evolutionary sense, and yet *where it would be wrong to act selfishly*. This explanation would also be really doing away with altruistic behavior because it is saying in effect that you should act altruistically because it will benefit you in the long run, but then it is not really altruistic behavior, because the essence of altruistic behavior is that you engage in it *without regard* to your own benefit in the short or long run.[36] Surely any evolutionary explanation that purports to explain altruism by showing how it benefits an individual or a group is not explaining *real* altruism, but actions that only appear to be (but are not) altruistic? Indirectly, these kinds of suggestions aimed at rescuing conventional morality for secularism also illustrate the power and depth of the Christian understanding of the moral order. It is difficult to avoid the conclusion that not only is the Christian moral view far superior to any moral theory founded on evolutionary naturalism, but, from the point of view of evolutionary theory, Christian moral behavior would have to be regarded as maladaptive because it promotes moral behavior that is often contrary to the "fitness" of the individual.

The significance of free will and moral agency

Our discussion about morality brings us finally to the question of human free will, and the fact that human beings are moral agents. It is in the context of the topic of evolution and morality that the phenomenon of free will is such a notoriously difficult problem for naturalistic theories of evolution, such as those advocated by Dawkins, Coyne, Mayr, Dennett, John Searle,[37] and others. I have indicated above that human free will would be outside the deterministic process of cause and effect I described as operating throughout the universe. Human free will is an indispensable concept in the religious view of the world, but it is also a central feature of any view of the world, even though we don't always focus on it explicitly in thinking about our moral, social, and political arrangements, but usually take it for granted. Free will may be defined as the ability of human beings to make a genuine choice between alternatives, a choice that is *not determined* by scientific laws operating on atomic or molecular particles or combinations of particles in the brain. Although there appears to be some cause and effect processes involved in free will, there also has to be an essential non-causal aspect to it, otherwise it would not be free will. On this understanding of free will, a scientific account of free will is not possible; it is a contradiction in terms. This is one clear area where the scientific approach runs up against its limits.

The decisions human beings make in the course of ordinary experience—like deciding what to eat for lunch—are arrived at freely in some mysterious way as part of the mental process of thinking, which involves thinking about the topic in hand, reasoning about it, then making a decision, and acting upon it. Yet, the reasons I have for making a decision do not causally compel me in the scientific sense to make my choice. Unlike, say, a machine that might be programmed to appear to be making a certain decision (the decision appears free but really is not), I am really free to make a decision. Free will is present in so much of what we do as human beings. I can begin thinking about various abstract problems, say the problem of evolution and religion, I can decide to go to college, to pursue a certain career, to take up bowling, or make countless other decisions; all of these decisions are based on my belief in and experience of free will. In fact, free will goes even deeper than this. For free will is at the foundations of morality, since the whole enterprise depends on human beings having a genuine choice between good and bad alternatives. Likewise, moral responsibility, punishment, and even democracy itself all depend on the prior belief that human beings have free will.[38]

Those committed to the doctrine of free will argue that there is no possible way to give a scientific account of free will because this is a contradiction in terms. It would be in effect asking for a scientific, causal account of something that is not subject to a causal explanation. What we can therefore say about free will is that it is real, but outside of physics, and beyond the scientific method. This is not a criticism of the scientific method, but a recognition that this method is limited to the physical realm, and that there are some things beyond the physical realm. This means that free will is the one area that is outside the determinism we talked about earlier. Human free will would not be totally subject to the laws of physics; and so human beings have the ability to intervene in causal chains and to alter the course of history. This is a very significant conclusion because not only does it mean that we are responsible for our own lives, but it also means that we can now influence to some extent some of the natural evil that occurs in the world, such as natural disasters and disease. It also means that we can control our own evolution ("direct" our own mutations) to some significant extent. So the fact that we can intervene in history is of great *moral* significance as well; perhaps this intervention is part of God's plan in order to contribute to our moral and spiritual development, as philosopher John Hick has argued.[39]

Ironically, some of the prominent evolutionary naturalists are often determinists at heart because determinism is required by their view that everything that exists, including the human mind and its activities, is physical in nature, and can be studied by science, at least in principle. But they forget that this kind of determinism must apply in the area of evolutionary change too, and so evolution cannot be random in any meaningful sense. In other words, they invite inconsistency because their rejection of free will (at least in their official view) requires a naturalistic, deterministic, and reductionistic account of reality; yet their view of chance operating in evolution contradicts this determinism. Many of these thinkers deal with the issue of free will by usually ignoring it, or by giving it inadequate attention (like Francis Crick's study of consciousness, in his book *The Astonishing Hypothesis*, which relegates the question of free will to a brief postscript[40]), despite the fact that it is a huge obstacle to a naturalistic account of human origins.

The notion of free will is related to other distinctive features of human nature, especially the nature of consciousness, and the human capacity for reason and logic. The question of consciousness is also connected to the question of the human soul. Many philosophers have considered human consciousness and the soul to be basically the same thing, and they have regarded the soul and the body as distinct entities; the soul is non-physical, and the body is physical, and the soul is what animates—in the language of Plato, Aristotle, St. Thomas Aquinas, and many others—the matter in the body. St. Thomas puts this by saying that "the soul is the principle of life."[41] The relationship between the soul and the mind is more complicated, and many traditional religious thinkers, in particular, did not fully work out their views on this topic. Some hold that the soul contains consciousness, others hold that consciousness is the same thing as the soul, but all are agreed that it is a remarkable quality of man, and that it cannot be explained in terms of, or reduced to, a physical explanation.[42]

It is significant that it is precisely the most significant features of human beings—consciousness, intelligence, free will, and moral agency—that have resisted scientific explanation, and that are extraordinary problems, not only for evolution, but also for naturalistic accounts of reality that rely on evolution (for these have to explain *all* facets of human existence, including free will, in scientific terms). This seems an impossible task. The alternatives are to say that free will is outside evolutionary explanation in the way we have explained, and so evolutionary naturalism collapses, or to say that we do not have free will (even though we think and act as if we do). This latter view is unthinkable and extremely counter-intuitive, to put it mildly. It is not rational to believe that human beings do not have free will, so any theory that argues that we do not have it, is not only impossible from a practical point of view, but also impossible from a theoretical point of view. We can give a similar analysis of other features of human existence such as consciousness, rationality, and logic, as well as the development of language and mathematics, to draw attention again to remarkable features of life that have emerged from the deterministic (and therefore teleological) process of evolution. It is a very significant point indeed that conscious, rational beings have evolved with free will that now have substantial control over the process of evolution. The enormity of this fact should not be lost on us. It points to the fact that something remarkable is going on, to the existence of an overall intelligence behind the processes in the universe.

It is also, finally, a powerful and convincing illustration of the compatibility of religion and evolution.

Notes

Chapter 1

1 For a spectrum of creationist views, see Richard Carlson (ed.), *Science and Christianity: Four Views* (Downers Grove, IL: InterVarsity, 2000), pp. 19–51; J.P. Moreland and John Mark Reynolds (eds.), *Three Views on Creation and Evolution* (Grand Rapids, MI: Zondervan, 1999); John Ashton (ed.), *In Six Days* (Green Forest, AZ: Master Books, 2001).

2 For Islamic thought, see S. H. Nasr, *The Need for a Sacred Science* (New York: SUNY Press, 1993); Martin Reixinger, "Responses of South Asian Muslims to the Theory of Evolution," *Die Welt Des Islams*, 49 (2009), pp. 212–247; for Indian thought, see B.V. Subbarayappa, "Indic Religions," in John H. Brooke and Ronald L. Numbers (eds.), *Science and Religion around the World* (New York: Oxford U.P., 2011).

3 For some representative works, see Richard Dawkins, *The Blind Watchmaker* (New York: Norton, 1996); Jerry Coyne, *Why Evolution Is True* (New York: Penguin, 2010); Carl Sagan, *Cosmos* (New York: Random House, 2002); William Provine, "Evolution and the Foundations of Ethics," *MBL Science*, 3 (1988), pp. 25–29; and Michael Ruse, *Darwin and Design: Does Evolution Have a Purpose*? (Cambridge, MA: Harvard U.P., 2004).

4 This general view is sometimes known by other names such as philosophical atheism, atheistic materialism, or by a more dated term, "scientism." A more common name today is "Darwinism," a term that was initially used to describe the cluster of ideas associated with Darwin's theory, but that later came to refer primarily to the secularist, atheistic philosophy that evolution was used to support (including theories of social Darwinism).

5 See Stephen Hawking and Leonard Mlodinow, *The Grand Design* (New York: Bantam, 2010), p. 8: "According to M-theory, ours is not the only universe. Instead, M-theory predicts that a great many universes were *created out of nothing*. Their creation does not require the intervention of some supernatural being or god. Rather, these multiple universes arise naturally from physical laws" (emphasis added).

6 Some representative works include, Paul Davies, *The Cosmic Blueprint* (New York: Simon and Schuster, 1988); Richard Swinburne, *Is There a God*? (New York: Oxford U.P., 1996); Ernan McMullin (ed.), *Evolution and Creation* (South Bend, IN: University of Notre Dame Press, 1985); Kenneth Miller, *Finding Darwin's God* (San Francisco, CA: Harper, 2007); Francis Collins, *The Language of God* (New York: Free Press, 2007); Keith Ward, *The Big Questions in Science and Religion* (West Conshohocken, PA: Templeton Press, 2008).

7 See R. Russell, W. Stoeger, and G. Coyne (eds.), *John Paul II on Science and Religion* (Vatican City: Vatican Observatory, 1990).
8 See Charles Hartshorne, *Omnipotence and Other Theological Mistakes* (New York: SUNY Press, 1983); see also John B. Cobb and David Ray Griffin, *Process Theology: An Introductory Exposition* (Philadelphia, PA: Westminster Press, 1976).
9 See John Polkinghorne, *Science and Theology* (Minneapolis, MA: Fortress, 1998); Ian Barbour, *When Science Meets Religion* (San Francisco, CA: Harper, 2000); John Haught, *God After Darwin* (Boulder, CO: Westview, 2000); Arthur Peacocke, *Theology for a Scientific Age* (Minneapolis, MN: Fortress, 1993); Philip Clayton, *God and Contemporary Science* (Grand Rapids, MI: Eerdman's, 1998).
10 See Nicholas Wolterstorff, *Practices of Belief* (New York: Cambridge, 2010), p. 345.
11 As an example, see Ian Barbour, *When Science Meets Religion*, pp. 174–180; also John Polkinghorne, *Science and Theology* (Minneapolis, MN: Fortress, 1998), ch. 2. I explore Polkinghorne's view in Chapter 7.
12 As an example, see Alex Rosenberg, *The Atheist's Guide to Reality* (New York: Norton, 2012).
13 See Geoffrey Bromiley, *Introduction to the Theology of Karl Barth* (Edinburgh: T. & T. Clark, 1979); Stephen J. Gould, *Rocks of Ages* (New York: Ballantine, 1999), ch. 1.
14 I owe this point to Bill Stancil.
15 For my own defense of these assumptions, see my *Religion: Key Concepts in Philosophy* (New York: Continuum, 2007); see also Richard Swinburne, *The Existence of God* (New York: Oxford, 2004), and John Haldane, *Reasonable Faith* (London: Routledge, 2010).

Chapter 2

1 See Arthur M. Silverstein, *Paul Ehrlich's Receptor Immunology* (London: Academic Press, 2001).
2 See Steven French, *Science: Key Concepts in Philosophy* (London: Continuum, 2007), ch. 1.
3 See Charles Darwin, *The Origin of Species* (London: Penguin, 1985 ed.); for a life of Darwin, see John Bowlby, *Charles Darwin* (New York: Norton, 1990).
4 See William Paley, *Natural Theology* (New York: Oxford, 2010 ed.).
5 See Charles Darwin, *Autobiography* (New York: Norton, 2005 ed.), p. 51; also p. 73.
6 See Charles Darwin, *The Origin of Species*, pp. 219, 458.
7 For more on the history of evolution, see Edward J. Larson, *Evolution: The Remarkable History of a Scientific Theory* (New York: Modern Library,

2006), especially ch. 3; also Peter J. Bowler, *Evolution: The History of an Idea* (Berkeley, CA: University of California Press, 1989 ed.).

8 The discussion in this section draws on my earlier book, *Religion and Science: An Introduction*, pp. 85–89.

9 Evolution is a difficult theory to approach for the first time. For excellent introductions, see Brian and Deborah Charlesworth, *Evolution: A Very Short Introduction* (New York: Oxford U.P., 2003); Stephen J. Gould, *Ever Since Darwin* (New York: Norton, 1992); Michael Ruse, *Darwinism and Its Discontents* (New York: Cambridge U.P., 2006); Jerry Coyne, *Why Evolution Is True*.

10 For an overview of human evolution, see Douglas Palmer, *Seven Million Years* (London: Phoenix Books, 2006).

11 See Charles Darwin, *The Origin of Species*, pp. 138ff.

12 See ibid., ch. 12 (esp. pp. 379ff).

13 Charles Darwin, *The Origin of Species*, p. 115.

14 See Christian De Duve, *Vital Dust* (New York: Basic Books, 1995), ch. 14; also Kenneth Miller, *Finding Darwin's God*, pp. 48–51.

15 See Charles Darwin, *The Origin of Species*, ch. IX, ch. XIII.

16 Ibid., pp. 454ff.

17 See Stephen J. Gould, *Wonderful Life* (New York: Norton, 1990), and Simon Conway Morris, *The Crucible of Creation* (New York: Oxford U.P., 2000), for contrasting accounts of this period.

18 This summary of evolutionary history is influenced by Kenneth Miller, *Finding Darwin's God*; see pp. 38ff.

19 See Carl Sagan, *Dragons of Eden* (New York: Ballantine, 1986), p. 14.

Chapter 3

1 See Charles Darwin, *The Origin of Species*, p. 133.

2 For critical reflections on natural selection, and other aspects of the theory, see Thomas Nagel, *Mind and Cosmos* (Oxford: Oxford U.P., 2012); Holmes Rolston, *Science and Religion* (Philadelphia, PA: Templeton, 1987), pp. 95–100; and Peter Van Inwagen, *God, Knowledge and Mystery* (Ithaca, NY: Cornell, 1995), pp. 128–162.

3 See Neil Campbell, *Biology* (Menlo Park, CA: Benjamin, 1996, 4th ed.), p. 408.

4 See Brian and Deborah Charlesworth, *Evolution: A Very Short Introduction*, pp. 79–89; also Jerry Coyne, *Why Evolution Is True*, ch. 5.

5 Stephen J. Gould and Richard Lewontin have argued, using their famous "spandrels" metaphor, that some features of an organism may not have adaptive value themselves but might be only "byproducts" of other features that have adaptive value, and so perhaps not every feature originates directly by means of natural selection; see their paper "The Spandrels of

San Marco and the Panglossian Paradigm: A Critique of the Adaptationist Program," in Steve Rose (ed.), *The Richness of Life: The Essential Stephen J. Gould* (New York: Norton, 2007), pp. 423–443.

6 For a good discussion of the overall topic, see Ernst Mayr, *Toward a New Philosophy of Biology* (Cambridge, MA: Harvard, 1988), pp. 402–422.

7 Charles Darwin, *The Origin of Species*, pp. 219–220.

8 Richard Dawkins, *The Blind Watchmaker*, p. 91.

9 For use of fictional examples, see Dawkins, ibid., pp. 77–109, p. 136, p. 240; also his *Climbing Mount Improbable* (New York: Norton, 1997), passim; also Jerry Coyne, *Why Evolution Is True*, p. 11; Christian De Duve, *Life Evolving*, p. 200.

10 See Ernst Mayr, *What Evolution Is* (New York: Basic, 2001), p. 275.

11 For a fascinating overview, see the discussion in Christian De Duve, *Vital Dust*, pp. 125–136.

12 Ibid., p. 126 (emphasis added).

13 For a good introduction, see David Ward, *Smithsonian Handbook: Fossils* (New York: DK, 2002).

14 See Sheridan Bowman, *Radiocarbon Dating* (Berkeley: University of California Press, 1990).

15 See Kennett Miller, *Finding Darwin's God*, p. 41.

16 See Charles Darwin, *The Origin of Species*, p. 458.

17 For an overview of the nature of DNA, see the exciting account in Francis Crick, *What Mad Pursuit* (New York: Basic Books, 1988); see also human genome research website, www.genome.gov

18 For more on the modern synthesis, see Peter Bowler, *Evolution: The History of an Idea*, pp. 307–332.

19 Francis Collins, *The Language of God*, p. 136.

20 Ibid., pp. 136–137.

21 Jerry Coyne, *Why Evolution Is True*, p. 222; for more on DNA evidence, see also Francisco Ayala, "The Theory of Evolution: Recent Successes," in Ernan McMullin (ed.), *Evolution and Creation*, pp. 59–90.

22 For an informative, objective discussion of the evidence for evolution, along with some of the religious implications of the theory, see Ernan McMullin, "Evolution and Special Creation," in *Zygon*, 28.3 (Sept. 1993), pp. 299–335.

23 See Ernst Mayr, *What Evolution Is*, ch. 11; Jerry Coyne, *Why Evolution Is True*, pp. 32–53; Stephen J. Gould, *The Panda's Thumb* (New York: Norton, 1990), pp. 181ff; Peter van Inwagen, *God, Knowledge and Mystery*, pp. 146–152; Samir Okasha, *Philosophy of Science* (London: Oxford U.P., 2002), pp. 128–129.

24 Charles Darwin, *The Origin of Species*, p. 207.

25 See Stephen J. Gould, "The Episodic Nature of Evolutionary Change," *The Panda's Thumb*, pp. 179–185; Gould and Niles Eldridge developed their theory of "punctuated equilibrium" partly as a response to this lack of fossil evidence; see Gould's *Punctuated Equilibrium* (Cambridge, MA: Belknap

Press, 2007). As an example of downplaying the problem, see Brian and Deborah Charlesworth, *Evolution: A Very Short Introduction*, p. 114.

26 See Jerry Coyne, *Why Evolution Is True*, p. 222; Stephen J. Gould, *Hen's Teeth and Horse's Toes* (New York: Norton, 1983), pp. 253–262; Ernst Mayr, *Toward a New Philosophy of Biology*, p. 192.

27 See Karl Popper, *Conjectures and Refutations* (London: Routledge, 2002), pp. 43–53.

28 See David Miller, *Popper Selections* (Princeton, NJ: Princeton, 1985), pp. 239–246.

29 See John Horgan, *The End of Science* (Indianapolis, IN: Addison-Wesley, 1996).

Chapter 4

1 See Alan Hirschfeld, *The Electric Life of Michael Faraday* (New York: Walker, 2006), pp. 161–173.

2 See Ronald L. Numbers, "Science without God: Natural Laws and Christian Beliefs," in David C. Lindberg and Ronald L. Numbers (eds.), *When Science and Christianity Meet* (Chicago: University of Chicago, 2003), pp. 265–285; see also Paul White, *Thomas Huxley* (New York: Cambridge, 2003).

3 See Daniel Dennett, *Darwin's Dangerous Idea* (New York: Simon & Schuster, 1995); see Richard Dawkins, *The Blind Watchmaker*.

4 For excellent discussions of the historical reaction to Darwin's theory, including the views of leading thinkers in both religion and science, see Jon Roberts, *Darwinism and the Divine in America* (Madison, WI: University of Wisconsin Press, 1988), and Ronald Numbers, *Darwinism Comes to America* (Cambridge, MA: Harvard U.P., 1998).

5 As quoted in Jon Roberts, *Darwinism and the Divine in America*, p. 22.

6 Ibid., p. 22.

7 See the fascinating, even-handed historical account in Ronald Numbers, *The Creationists* (Berkeley, CA: University of California Press, 1993).

8 See Keith Ward, *God, Chance and Necessity* (Oxford: Oneworld, 1996); Kenneth Miller, *Finding Darwin's God*; Paul Davies, *The Cosmic Blueprint*; Richard Swinburne, *Is There a God*? (New York: Oxford U.P., 1996).

9 See Ian Barbour, *Religion and Science* (San Francisco, CA: Harper, 1997); Arthur Peacocke, *Theology for a Scientific Age*.

10 See Alvin Plantinga "When Faith and Reason Clash: Evolution and the Bible," in *Christian Scholar's Review*, XXI.1 (Sept. 1991), p. 17.

11 Daniel Dennett, *Darwin's Dangerous Idea*, p. 203.

12 Alvin Plantinga, "When Faith and Reason Clash," p.17.

13 See Thomas Nagel, *Mind and Cosmos*, and Jerry Fodor and Massimo Piattelli-Palmarini, *What Darwin Got Wrong* (New York: Farrar, 2010). For a

summary of the reaction, see Mark Vernon, "The Most Despised Science Book of 2012 Is Worth Reading," *The Guardian*, Friday, Jan. 4, 2013.

14 For a good illustration of Dennett's attitude, see his debate with Alvin Plantinga, *Science and Religion: Are They Compatible*? (New York: Oxford U.P., 2011).

15 For an example, see Burton Guttman, *Evolution: A Beginner's Guide* (Oxford: Oneworld, 2005); one that crosses the line into advocacy is Jay Phelan, *What Is Life? A Guide to Biology* (New York: Freeman, 2010).

16 Many biology textbooks now include the subject of the origin of life as a part of the theory of evolution. As examples, see Neil Campbell, *Biology*; Kenneth Miller and Joseph Levine, *Biology* (Englewood Cliffs, NJ: Prentice Hall, 1995 ed.); William Schraer and Herbert Stoltze, *Biology* (Hoboken, NJ: Pearson, 1999).

17 See Simon Conway Morris, *Life's Solution* (New York: Cambridge U.P., 2003), pp. 44–68 for a full discussion; also Christian de Duve, *Life Evolving*, chs. 4–8.

18 See Carl Sagan, *Cosmos*, ch. 8; also Eric Chaisson, *Cosmic Evolution* (Cambridge, MA: Harvard U.P., 2002).

19 For a perceptive critique of these views in a number of influential thinkers, see Karl Giberson and Mariano Artigas, *Oracles of Science* (New York: Oxford U.P., 2007).

20 Brian and Deborah Charlesworth, *Evolution*, pp. 126–127. For a more even treatment of questions that are very problematic for naturalism, see Samir Okasha, *Philosophy of Science*, pp. 52–57.

Chapter 5

1 For discussions of chance and randomness in evolution by contemporary evolutionary biologists, see Richard Dawkins, *The Blind Watchmaker*, pp. 306ff; Jerry Coyne, *Why Evolution Is True*, pp. 118–124; Jacques Monod, *Chance and Necessity* (New York: Knopf, 1971), passim; Ernst Mayr, *What Evolution Is*, pp. 119–124.

2 Stephen J. Gould, *Wonderful Life*, p. 51; see also pp. 309ff. Other thinkers in both science and religion who hold that evolution operates by chance include Kenneth Miller, *Finding Darwin's God*, pp. 232–239; Francisco Ayala, "Design without Designer: Darwin's Greatest Discovery," in William Dembski and Michael Ruse (eds.), *Debating Design* (New York: Cambridge U.P., 2004), pp. 55–80; John Haught, *Science and Faith* (Mahwah, NJ: Paulist, 2012), pp. 46ff; Arthur Peacocke, *Theology for a Scientific Age*, pp. 64–65.

3 See Richard Dawkins, *The Blind Watchmaker*, pp. 306ff.

4 Ibid., p. 306.

5 See Elliott Sober, "Evolution with Naturalism," in Jonathan Kvanvig (ed.), *Oxford Studies in the Philosophy of Religion*, Vol. 3 (New York: Oxford U.P., 2011), p. 191 (emphasis added).

6 See Jacques Monod, *Chance and Necessity*, pp. 112–113.

7 For some general discussions of chance, and related concepts, see Keith Ward, *God, Chance and Necessity*; David Ruelle, *Chance and Chaos* (Princeton, NJ: Princeton, 1991); David Bartholomew, *God, Chance and Purpose* (Cambridge: Cambridge U.P., 2008).

8 For an overview of various approaches to teleology and chance, see Ernan McMullin, "Cosmic Purpose and the Contingency of Human Evolution," *Zygon*, 48.2 (June 2013), pp. 338–363. See also Neil Manson (ed.), *God and Design* (New York: Routledge, 2003); Edward Feser, "Teleology: A Shopper's Guide," *Philosophia Christi*, 12.1 (2010), pp. 142–159.

9 For a speculative, but interesting, discussion of the origin and development of the early universe, see Steven Weinberg, *The First Three Minutes* (New York: Basic Books, 1993).

10 For general discussions of determinism, indeterminism, and chance, see Paul Humphreys, *The Chances of Explanation* (Princeton, NJ: Princeton U.P., 2014); Karl Popper, *The Open Universe* (Routledge: London, 1982); Stuart Glennan, "Probable Causes and the Distinction between Subjective and Objective Chance," *Nous*, 31.4 (1997), pp. 486–519.

11 See Pierre Laplace, *A Philosophical Essay on Probabilities* (Hong Kong: Forgotten Books, 2012 ed.).

12 The theologian, Arthur Peacocke, suggests that when causal chains interact with each other, they do so by chance; see his *Theology for a Scientific Age*, p. 64. (I discuss Peacocke's view in Chapters 7 and 8.) This view also appears to be held by Philip Clayton, see his *God and Contemporary Science*, ch. 7. It seems also to have been the view of Aristotle, though the examples he uses to illustrate involve the interaction of human beings (which would not be determined, on my view), rather than examples from the natural world of cause and effect (see Aristotle, *Physics*, Book II); see also John Dudley, *Aristotle's Concept of Chance* (Albany, NY: SUNY Press, 2012).

Chapter 6

1 See Jerry Coyne, *Why Evolution Is True*, p. 118.

2 See Francisco Ayala, "Design without Designer: Darwin's Greatest Discovery," in William Dembski and Michael Ruse (eds.), *Debating Design*, pp. 65–67.

3 See the following two representative textbooks widely used in schools and colleges that present evolution as operating by chance: Peter Raven, *et al.*, *Biology* (New York: McGraw Hill, 2008, 8th ed.), pp. 401ff; and Jay Phelan, *What Is Life?*, pp. 260–264. Phelan confuses and conflates probability with chance in nature throughout his exposition.

4 See Richard Dawkins, *Climbing Mount Improbable*, p. 79.

5 Ernst Mayr, *What Evolution Is*, p. 119.

6 Ibid., p. 120.

7 Ibid., p. 121.
8 Stephen J. Gould, *Wonderful Life*, pp. 320–321 (emphasis added).
9 Ibid., p. 309.
10 For an overview from the point of view of paleontology, see George Gaylord Simpson, *This View of Life* (Fort Worth, TX: Harcourt, 1966), pp. 176–189.
11 Simon Conway Morris, *Life's Solution*, p. 282.
12 There is quite a debate within biology about whether there may be a "directed" process of mutation at work at the molecular level; for an overview of the debate, see Richard Lenski and John Miller, "The Directed Mutation Controversy and Neo-Darwinism," *Science*, 259 (Jan. 1993), pp. 188–194. One would be unaware that there is such a debate from a reading of most biology textbooks.
13 Coyne argues for chance in nature when discussing evolution, but for determinism when discussing free will; see Jerry Coyne, "You Don't Have Free Will," *The Chronicle of Higher Education*, Mar. 18, 2012.
14 For a good overview of the positions of several thinkers, including discussion of various examples, see Roberta L. Millstein, "How Not to Argue for the Indeterminism of Evolution," in Andreas Huttemann (ed.) *Determinism in Physics and Biology* (Paderborn: Mentis, 2003), pp. 91–107.
15 Jerry Coyne, *Why Evolution Is True*, pp. 122–123 (emphasis added).
16 For an introduction to probability, and related notions, see D.H. Mellor, *Probability: A Philosophical Introduction* (London: Routledge, 2005).
17 See Pierre Laplace, *A Philosophical Essay on Probabilities*, p. 4.
18 Jacques Monod, *Chance and Necessity*, p. 114 (parenthetical remark added).
19 Jerry Coyne, *Why Evolution Is True*, p. 122 (emphasis added).
20 See Jacques Monod, *Chance and Necessity*, pp. 123ff.
21 Charles Darwin, *The Origin of Species*, p. 123.
22 For a good overview of quantum theory, including the various interpretative problems, see John Polkinghorne, *Quantum Theory* (New York: Oxford U.P., 2002).
23 See David Bohm, *Quantum Theory* (New York: Dover, 1989). See Einstein's letter to Max Born in *The Born–Einstein Letters*, trans. Irene Born (New York: Walker, 1971), pp. xxii.
24 See a very helpful article by Shanahan, "The Evolutionary Indeterminism Thesis," *BioScience*, 53.2 (Feb. 2003), pp. 163–169; see also, William Lane Craig, "The Caused Beginning of the Universe: A Response to Quentin Smith," *British Journal for the Philosophy of Science* 44 (1993), pp. 623–639; see also, Alex Rosenburg, "Is the Theory of Natural Selection a Statistical Theory," *Canadian Journal of Philosophy*, 14 (1988), pp. 187–207.
25 See Charles Darwin, *The Origin of Species*, p. 459; for an overview of Darwin's and of contemporary views, see Michael Ruse, *Monad to Man* (Cambridge, MA: Harvard U.P., 1997), pp. 150–169; pp. 485–525. See also Ernst Mayr, *What Evolution Is*, pp. 212ff; also p. 278; Christian de Duve, *Vital Dust*, pp. 292ff; Paul Davies, *The Cosmic Blueprint*, pp. 107–120.

Chapter 7

1 See the embarrassingly weak attempt to address the question of the ultimate cause of the universe in Richard Dawkins, *The God Delusion* (New York: Houghton Mifflin, 2006), pp. 77–79; Daniel Dennett's misstatement of the cosmological argument for God's existence in *Breaking the Spell* (New York: Viking, 2006), p. 242; also Francis Crick, *The Astonishing Hypothesis* (New York: Touchstone, 1995), pp. 5ff, where he lays out his naturalistic thesis but does not mention the question of the origin of the universe.

2 For a full discussion, see Richard Swinburne, *The Existence of God*; William Lane Craig, *The Cosmological Argument from Plato to Leibniz* (Eugene, OR: Wipf & Stock, 2001); Brendan Sweetman, *Religion: Key Concepts in Philosophy.*

3 William Rowe, *Philosophy of Religion* (Belmont, CA: Wadsworth, 2007, 4th ed.), p. 32.

4 See Paul Edwards, "A Critique of the Cosmological Argument," in L. Pojman (ed.), *Philosophy of Religion: An Anthology* (Belmont, CA: Wadsworth, 2003), pp. 59–73; see Bertrand Russell, *Why I Am Not a Christian* (London: George Allen & Unwin, 1957), pp. 133–153; see Paul Draper, "A Critique of the Kalam Cosmological Argument," in the Pojman anthology, pp. 42–47.

5 The questions of the ultimate origin of the universe and of life, and usually also of the laws of the universe, are routinely ignored by atheistic naturalists: as examples, see Richard Dawkins, *The God Delusion*, a book that has no discussion of the question of the origin of the universe or of the origin of the laws of physics. Michael Shermer's, *How We Believe: Science, Skepticism and the Search for God* (New York: Holt, 2000) is another example; yet another is Daniel Dennett's, *Darwin's Dangerous Idea*, a book that purports to be explaining the meaning of life in all its aspects! One lone naturalist who does acknowledge the failure of naturalism to offer arguments with regard to the ultimate questions is Quentin Smith, who admits that: "the great majority of naturalist philosophers have an unjustified belief that naturalism is true and an unjustified belief that theism (or supernaturalism) is false"; see his "The Metaphilosophy of Naturalism," *Philo*, 4.2 (2001), p. 195. One naturalist who does address the ultimate question is Peter Atkins, who concludes that: "… spacetime generates its own dust in the process of its own self assembly. The universe can emerge out of nothing. By chance."; see his *The Creation* (London: Oxford U.P., 1981), p. 113.

6 For a penetrating critical discussion of naturalism, see Stewart Goetz and Charles Taliaferro, *Naturalism* (Grand Rapids, MI: Eerdmans, 2008).

7 See St. Thomas Aquinas, *Summa Theologiae*, Part 1, Question 2, Article 3, in Anton Pegis (ed.), *Introduction to St. Thomas Aquinas* (New York: Random House, 1945), pp. 24–27; see also *Summa Contra Gentiles*, Part 1, chapters 9–14, in Ralph McInerny (ed.), *Thomas Aquinas: Selected Writings*, (Middlesex: Penguin, 1998), pp. 243–256; also Frederick Copleston, *Aquinas* (Harmondsworth: Pelican, 1975), pp. 121ff; and Richard Taylor, *Metaphysics* (Englewood Cliffs, NJ: Prentice Hall, 1963).

8 See Dallas Willard, "The Three-Stage Argument for the Existence of God," in R. Douglas Geivett and Brendan Sweetman (eds.), *Contemporary Perspectives on Religious Epistemology* (New York: Oxford U.P., 1992), p. 216.
9 See also Sagan's book *Cosmos*, ch. 1.
10 Ibid., pp. 30–31, where he presents this speculative view as if it were fact; on p. 39, he is more circumspect in what he claims for the abiogenesis theory.
11 In addition to those mentioned in previous chapters, see Kenneth Miller's *Finding Darwin's God*, p. 236, where he confuses chance with unpredictability, and also conspicuously fails to distinguish between causal processes in nature and events that are brought about by human free will (despite his very good discussion of other major issues in the book); see also Christian de Duve, *Vital Dust*, who argues that "life is the product of deterministic forces" (p. xvii), but who then seems to contradict himself by claiming that there is "contingency" involved in the process (xvi). See also John Beatty, "Replaying Life's Tape," *The Journal of Philosophy* 103.7 (Jul. 2006), pp. 336–362; Beatty does not provide clear definitions of chance, randomness, historical contingency, or predictability, and as a result confuses and conflates these notions; see also Daniel Dennett, *Darwin's Dangerous Idea*, p. 112, where he confuses chance with unpredictability when discussing changes in the human genome. The literature on evolution is full of these kinds of confusions and conflations.
12 Carl Sagan, *Cosmos*, p. 246.
13 Carl Sagan, *Billions and Billions* (New York: Ballantine, 1997), pp. 62–63.
14 For a very helpful overview and critique of this view, see Alvin Plantinga, *Where the Conflict Really Lies* (New York: Oxford U.P., 2011), pp. 97–113. See also the excellent general discussion in Nicholas Saunders, *Divine Action and Modern Science* (Cambridge: Cambridge U.P., 2002).
15 See George F. R. Ellis, "Ordinary and Extraordinary Divine Action," in Robert Russell, *et al.* (eds.), *Chaos and Complexity* (Vatican City: Vatican Observatory, 2000), pp. 359–395.
16 See Ellis essay, ibid.; also Polkinghorne's essay, "The Metaphysics of Divine Action," pp. 147–156 in the same volume. Also, John Haught, *Science and Faith* (Mahwah, NJ: Paulist, 2012), ch. 4.
17 Some theologians hold that God could intervene at the quantum level in a way that is undetectable to science; see Nancey Murphy, *Beyond Liberalism and Fundamentalism* (Valley Forge, PA: Trinity Press, 1996), pp. 340ff.
18 For an excellent overview of this distinction, and of St. Thomas's view of causation, see Michael Dodds, "The Doctrine of Causality in Aquinas and *The Book of Causes*: One Key to Understanding the Nature of Divine Action," in Timothy L. Smith (ed.), *Aquinas's Sources* (South Bend, IN: St. Augustine's Press, 2015); see also Alfred Freddoso, "Medieval Aristotelianism and the Case against Secondary Causation in Nature," in Thomas Morris (ed.), *Divine and Human Action* (Ithaca, NY: Cornell, 1988), pp. 74–118.
19 For an interesting discussion, see William R. Stoeger, S.J., "Describing God's Action in the World in Light of Scientific Knowledge of Reality," in Robert Russell, *et al.*, *Chaos and Complexity*, pp. 239–261.

20 See my *Religion and Science: An Introduction*, pp. 104–106; see also Ernan McMullin (ed.), *Evolution and Creation*, pp. 11–16, and his "Evolution as a Christian Theme," Reynolds Lecture, Baylor University, Mar. 2004 (pdf. available at www.baylor.edu); also William Carroll, "Creation, Evolution and Thomas Aquinas," *Revue des Questions Scientifiques* 171.4 (2000), pp. 319–347.
21 See R. Russell, *et al.* (eds.), *John Paul II on Science and Religion*; Christoph Cardinal Schőnborn, *Chance or Purpose?* (San Francisco, CA: Ignatius, 2007); Don O'Leary, *Roman Catholicism and Modern Science* (London: Continuum, 2007); see also the essays in Robert Russell, *et al.*, *Evolutionary and Molecular Biology* (Vatican City: Vatican Observatory, 1998).
22 The importance of truth—our beliefs matching up with reality—should not be underestimated. See Plantinga's intriguing argument that it would be improbable that human beings could have true beliefs if our cognitive faculties had come about by means of an unplanned evolutionary process, and this suggests that *naturalism* and evolution are, in fact, incompatible; see his *Where the Conflict Really Lies*, pp. 307–350.
23 Alfred R. Wallace, "The Limits of Natural Selection as Applied to Man," in his *Contributions to the Theory of Natural Selection* (London: Macmillan, 1870), p. 221.
24 See Keith Ward, *God, Chance and Necessity*, pp. 64–69.
25 This view is discussed in E.L. Mascall, *Christian Theology and Natural Science* (New York: Ronald, 1956), pp. 311–316.
26 One thinker who argues that it is *improbable* that God would intervene in nature in acts of special creation is Ernan McMullin; see his "Evolution and Special Creation," *Zygon*, 28.3 (Sept. 1993), pp. 323–326; see also McMullin's Augustinian account of creation in "Cosmic Purpose and the Contingency of Human Evolution," *Zygon*, 48.2 (June 2013), pp. 338–363.
27 See Michael Behe, *Darwin's Black Box: The Biochemical Challenge to Evolution* (New York: Free Press, 1998).
28 John Polkinghorne, *Science and Theology*, p. 39.
29 Ibid., p. 39.
30 Ibid., p. 79.
31 Ibid., p. 79.
32 See Arthur Peacocke, *Theology for a Scientific Age*, pp. 49ff.
33 Ibid., pp. 64–65; also p. 115.
34 Ibid., p. 65.

Chapter 8

1 Including, apparently, Ernan McMullin; see his "Evolution as a Christian Theme," section 8, Reynolds Lecture, Baylor University, Mar. 2004 (pdf, available at www.baylor.edu).

2 See Steven Weinberg, *The First Three Minutes*.

3 See Arthur Peacocke, *Theology for a Scientific Age*, pp. 118–119; for Dawkins's discussion of the computer models, see *The Blind Watchmaker*, ch. 3.

4 See Ernan McMullin, "Evolution and Special Creation," *Zygon*, 28.3 (Sept. 1993), p. 324.

5 See Richard Dawkins, *The Blind Watchmaker*, pp. 43–44.

6 See John Barrow and Frank Tipler, *The Anthropic Cosmological Principle* (New York: Oxford, 1988); see also an excellent set of essays in Neil Manson (ed.), *God and Design*.

7 St. Thomas Aquinas, *On Truth* (*Questiones Disputatae de Veritate*), Q. 5, art. 2, in any edition.

8 Dallas Willard, "The Three-Stage Argument for the Existence of God," in R. Douglas Geivett and Brendan Sweetman (eds.), *Contemporary Perspectives on Religious Epistemology*, p. 218.

9 See Alvin Plantinga, "When Faith and Reason Clash: Evolution and the Bible," *Christian Scholar's Review*, XXI.1 (Sept. 1991), pp. 16–19.

10 See Richard Dawkins, *The Blind Watchmaker*, p. 6.

11 See Ernan McMullin, "Plantinga's Defense of Special Creation," *Christian Scholar's Review*, XXI.1 (Sept. 1991), p. 55.

12 See Michael Behe, *Darwin's Black Box*, and *The Edge of Evolution* (New York: Free Press, 2007); William Dembski, *The Design Revolution* (Downers Grove, IL: InterVarsity, 2004); Phillip Johnson, *Darwin on Trial* (Downers Grove, IL: InterVarsity, 1993).

13 See Coyne's screed against religion, "Science and Religion Aren't Friends," *USA Today*, Oct. 11, 2010.

14 ID theorists occasionally suggest that God could design the plan of living things, including the irreducible complexities, into the blueprint of the universe from the beginning of creation, and then there would be no need for God's direct intervention in biological systems in acts of special creation. However, this claim is contrary to the way ID theory is normally presented and would seem to make it indistinguishable from more traditional forms of the argument from design, especially if ID theorists then add that only *random* natural selection could not produce biological complexity, but that *directed* (i.e., non-random) natural selection *could* do so. If they nuance their view in this way, then their only original claim would appear to be that ID is part of science, not philosophy or theology. Their other claims would be indistinguishable from theistic evolution, a view they reject. See William Dembski, *No Free Lunch* (New York: Rowman & Littlefield, 2002), p. 335; Michael Behe, *Darwin's Black Box*, pp. 229–233.

15 For the distinction between methodological and metaphysical naturalism, see Paul de Vries, "Naturalism in the Natural Sciences: A Christian Perspective," *Christian Scholar's Review*, 15.4 (1986), pp. 388–396.

16 For an excellent discussion of the issues surrounding our inability to make predictions based on evolutionary theory, see Bruce Reichenbach, "Justifying

In-Principle Nonpredictive Theories: The Case of Evolution," *Christian Scholar's Review*, XXIV.4 (May 1995), pp. 397–422.

17 For a critique of ID, see Kenneth Miller, *Finding Darwin's God*; Robert Pennock (ed.), *Intelligent Design Creationism and Its Critics* (New York: MIT Press, 2001); Elliott Sober, *Evidence and Evolution* (Cambridge: Cambridge U.P., 2008), ch. 2.

18 For some examples, see Kenneth Miller, ibid.; Richard Dawkins, *The Blind Watchmaker*, pp. 77–109; also p. 136, p. 240; and Christian De Duve, *Life Evolving*, p. 200.

19 For a discussion of the problem of evil, see Michael Peterson, *God and Evil* (Boulder, CO: Westview, 1988); James Petrik, *Evil beyond Belief* (Armonk, NY: ME Sharpe, 2000); Brendan Sweetman, *Religion: Key Concepts in Philosophy* (New York: Continuum, 2007), ch. 3.

20 For a discussion of this example, and historical reaction to it, see Stephen J. Gould, *Hen's Teeth and Horse's Toes*, pp. 32–45. See also Philip Kitcher, *Living with Darwin* (New York: Oxford, 2007), pp. 123ff.

21 On the evolution of the eye, see Brian and Deborah Charlesworth, *Evolution: A Very Short Introduction*, pp. 115–117; Christian De Duve, *Life Evolving*, p. 200; Burton Guttman, *Evolution: A Beginner's Guide*, pp. 10–11.

22 See Burton Guttman, ibid., pp. 10–11.

23 See Stephen J. Gould, *The Panda's Thumb*.

24 See Richard Dawkins, *The Blind Watchmaker*, p. 306.

25 See Stephen J. Gould, *Rocks of Ages*, pp. 198–207.

26 For some responses to the problem of evil, see St. Augustine, *City of God*, xii, sections 6–7; C.S. Lewis, *The Problem of Pain* (New York: Macmillan, 1962); John Hick, *Evil and the God of Love* (New York: Harper and Row, 1966); and a recent excellent study, Michael Murray, *Nature Red in Tooth and Claw* (New York: Oxford, 2011).

27 See Keith Ward, *God, Chance and Necessity*, pp. 80–95; Francisco Ayala, *Darwin's Gift to Science and Religion* (Washington, DC: Joseph Henry, 2007), pp. 159–160.

28 See Immanuel Kant, *Critique of Practical Reason*, trans. W.S. Pluhar (Indianapolis, IN: Hackett, 2002), p. 115.

29 See the excellent discussion in Michael Devitt, "Resurrecting Biological Essentialism," *Philosophy of Science*, 75 (July 2008), pp. 344–382.

30 See Edward O. Wilson, *Sociobiology: The New Synthesis* (Cambridge, MA: Harvard U.P., 1975); for a critique of this view, see S. Rose, *et al.*, *Not in Our Genes* (London: Penguin, 1990); and Stephen J. Gould, "Sociobiology and the Theory of Natural Selection," in George Barlow and James Silverberg (eds.), *Sociobiology: Beyond Nature/Nurture*? (Boulder, CO: Westview Press, 1980), pp. 257–269.

31 See Michael Ruse, *Taking Darwin Seriously* (Amherst, NY: Prometheus, 1998), p. 253.

32 See Richard Dawkins, *River Out of Eden* (New York: Basic, 1996), pp. 131–133.

33 See Richard Dawkins, "Let's All Stop Beating Basil's Car," at *Edge: The World Question Center*, at http://edge.org/q2006/q06_9.html#dawkins
34 See Richard Dawkins, *The God Delusion*, ch. 5, and Dennett, *Breaking the Spell*, passim.
35 For a history of this movement, see Daniel Kevles, *In the Name of Eugenics* (Los Angeles: University of California Press, 1985); see also some disturbing passages in Charles Darwin, *The Descent of Man* (London: Penguin 2004 ed.), p. 159.
36 See Larry Arnhart, *Darwinian Natural Right* (New York: SUNY Press, 1998), for an alternative perspective on some of these issues; also Samir Okasha, *Evolution and the Levels of Selection* (Oxford: Oxford U.P., 2006).
37 See John Searle, *Freedom and Neurobiology* (New York: Columbia U.P., 2006), pp. 37–78.
38 Jerry Coyne holds that moral responsibility will have to be discarded, along with our belief in free will; see his "You Don't Have Free Will," *The Chronicle of Higher Education*, Mar. 18, 2012.
39 See John Hick, *Evil and the God of Love* (New York: Harper and Row, 1966).
40 See Francis Crick, *The Astonishing Hypothesis*, pp. 265–268.
41 St. Thomas Aquinas, *Summa Theologiae*, Part II, II, Q.179; also Ralph McInerny (ed.), *Thomas Aquinas: Selected Writings*, pp. 410–428.
42 For further discussion, see St. Thomas Aquinas, *Questions on the Soul*, trans. James Robb (Milwaukee: Marquette U.P., 1984); Joel Green and Stuart Palmer (eds.), *In Search of the Soul* (Downers Grove, IL: InterVarsity, 2005); and Nancey Murphy, *Bodies and Souls, or Spirited Bodies?* (New York: Cambridge U.P., 2006).

Further Reading

Aquinas, Thomas, *Selected Writings*, ed. by Ralph McInerny (New York: Penguin, 1998).

Arnhart, Larry, *Darwinian Natural Right* (New York: SUNY Press, 1998).

Artigas, Mariano, *The Mind of the Universe* (Philadelphia, PA: Templeton Press, 2000).

Ashton, John (ed.), *In Six Days* (Green Forest, AZ: Master Books, 2001).

Atkins, Peter, *The Creation* (London: Oxford U.P., 1981).

Augustine, St., *City of God* (London: Penguin, 2003).

Augustine, St., *On Genesis*, ed. by B. Ramsey (New York: New City Press, 2004).

Ayala, Francisco, *Darwin's Gift to Science and Religion* (Washington, DC: Joseph Henry, 2007).

Barbour, Ian, *When Science Meets Religion* (San Francisco, CA: Harper, 2000).

Barbour, Ian, *Religion and Science* (San Francisco: Harper, 1997).

Barlow, George and James Silverberg (eds.), *Sociobiology: Beyond Nature/Nurture?* (Boulder, CO: Westview Press, 1980).

Barrow, John and Frank Tipler, *The Anthropic Cosmological Principle* (New York: Oxford U.P., 1988).

Bartholomew, David, *God, Chance and Purpose* (Cambridge: Cambridge U.P., 2008).

Beatty, John, "Replaying Life's Tape," *The Journal of Philosophy* 103.7 (July 2006), pp. 336–362.

Behe, Michael, *Darwin's Black Box: The Biochemical Challenge to Evolution* (New York: Free Press, 1998).

Behe, Michael, *The Edge of Evolution* (New York: Free Press, 2007).

Bohm, David, *Quantum Theory* (New York: Dover, 1989).

Bowlby, John, *Charles Darwin: A New Life* (New York: Norton, 1990).

Bowler, Peter, *Evolution: The History of an Idea* (Berkeley, CA: University of California Press, 1989 ed.).

Brooke, John and Ronald Numbers (eds.), *Science and Religion around the World* (New York: Oxford U.P., 2011).

Carlson, Richard (ed.), *Science and Christianity: Four Views* (Downers Grove, IL: InterVarsity, 2000).

Carroll, William, "Creation, Evolution and Thomas Aquinas," *Revue des Questions Scientifiques* 171.4 (2000), pp. 319–347.

Chaisson, Eric, *Cosmic Evolution* (Cambridge, MA: Harvard U.P., 2002).

Charlesworth, Brian and Deborah, *Evolution: A Very Short Introduction* (New York: Oxford U.P., 2003).

Clayton, Philip, *God and Contemporary Science* (Grand Rapids, MI: Eerdman's, 1998).

Cobb, John and David Ray Griffin, *Process Theology: An Introductory Exposition* (Philadelphia, PA: Westminster Press, 1976).
Collins, Francis, *The Language of God* (New York: Free Press, 2007).
Coyne, Jerry, *Why Evolution Is True* (New York: Penguin, 2010).
Coyne, Jerry, "Science and Religion Aren't Friends," *USA Today*, Oct. 11, 2010.
Coyne, Jerry, "You Don't Have Free Will," *The Chronicle of Higher Education*, Mar. 18, 2012.
Craig, William Lane, *The Cosmological Argument from Plato to Leibniz* (Eugene, OR: Wipf & Stock, 2001).
Crick, Francis, *The Astonishing Hypothesis* (New York: Touchstone, 1995).
Darwin, Charles, *The Autobiography of Charles Darwin 1809–1882* (New York: Norton, 2005).
Darwin, Charles, *The Descent of Man* (London: Penguin, 2004 ed.).
Darwin, Charles, *The Origin of Species* (London: Penguin, 1985 ed.).
Davies, Paul, *The Cosmic Blueprint* (New York: Simon and Schuster, 1988).
Dawkins, Richard, *The Blind Watchmaker* (New York: Norton, 1996).
Dawkins, Richard, *Climbing Mount Improbable* (New York: Norton, 1997).
Dawkins, Richard, *The God Delusion* (New York: Houghton Mifflin, 2006).
Dawkins, Richard, *River Out of Eden* (New York: Basic, 1996).
De Duve, Christian, *Vital Dust* (New York: Basic Books, 1995).
De Vries, Paul, "Naturalism in the Natural Sciences: A Christian Perspective," *Christian Scholar's Review*, 15.4 (1986), pp. 388–396.
Dembski, William, *The Design Revolution* (Downers Grove, IL: InterVarsity, 2004).
Dembski, William, *No Free Lunch: Why Specified Complexity Cannot Be Purchased without Intelligence* (New York: Rowman & Littlefield, 2002).
Dembski, William and Michael Ruse (eds.), *Debating Design: From Darwin to DNA* (New York: Cambridge, 2004).
Dennett, Daniel, *Breaking the Spell* (New York: Viking, 2006).
Dennett, Daniel, *Darwin's Dangerous Idea* (New York: Simon & Schuster, 1995).
Dennett, Daniel and Alvin Plantinga, *Science and Religion: Are They Compatible?* (New York: Oxford U.P., 2011).
Denton, Michael, *Nature's Destiny: How the Laws of Biology Reveal Purpose in the Universe* (New York: Free Press, 1998).
Dudley, John, *Aristotle's Concept of Chance* (Albany, NY: SUNY Press, 2012).
Fodor, Jerry and Massimo Piattelli-Palmarini, *What Darwin Got Wrong* (New York: Farrar, 2010).
Geivett, R. Douglas and Brendan Sweetman (eds.), *Contemporary Perspectives on Religious Epistemology* (New York: Oxford U.P., 1992).
Giberson, Karl and Mariano Artigas, *Oracles of Science* (New York: Oxford U.P., 2007).
Gilson, Etienne, *From Aristotle to Darwin and Back Again* (San Francisco, CA: Ignatius, 2009).
Gingerich, Owen, *God's Universe* (Cambridge, MA: Harvard U.P., 2006).

Glennan, Stuart, "Probable Causes and the Distinction between Subjective and Objective Chance," *Nous*, 31.4 (1997), pp. 486–519.
Goetz, Stewart and Charles Taliaferro, *Naturalism* (Grand Rapids, MI: Eerdmans, 2008).
Gould, Stephen J., *Ever Since Darwin* (New York: Norton, 1992).
Gould, Stephen J., *The Panda's Thumb* (New York: Norton, 1990).
Gould, Stephen J., *Rocks of Ages* (New York: Ballantine, 1999).
Gould, Stephen J., *Wonderful Life: The Burgess Shale and the Nature of History* (New York: Norton, 1990).
Green, Joel and Stuart Palmer (eds.), *In Search of the Soul: Four Views of the Mind-Body Problem* (Downers Grove, IL: InterVarsity, 2005).
Guttman, Burton, *Evolution: A Beginner's Guide* (Oxford: Oneworld, 2005).
Haldane, John, *Reasonable Faith* (London: Routledge, 2010).
Hartshorne, Charles, *Omnipotence and Other Theological Mistakes* (New York: SUNY Press, 1983).
Haught, John, *God After Darwin* (Boulder, CO: Westview, 2000).
Haught, John, *Science and Faith* (Mahwah, NJ: Paulist, 2012).
Hawking, Stephen and Leonard Mlodinow, *The Grand Design* (New York: Bantam, 2010).
Heisenberg, Werner, *Physics and Philosophy* (New York: Prometheus, 1999).
Hick, John, *Evil and the God of Love* (New York: Harper and Row, 1966).
Humphreys, Paul, *The Chances of Explanation* (Princeton, NJ: Princeton U.P., 2014).
Johnson, Phillip, *Darwin on Trial* (Downers Grove, IL: InterVarsity, 1993).
Kane, Robert, *A Contemporary Introduction to Free Will* (New York: Oxford, 2005).
Kevles, Daniel, *In the Name of Eugenics* (Los Angeles, CA: University of California Press, 1985).
Kitcher, Philip, *Living with Darwin* (New York: Oxford U.P., 2007).
Kvanvig, Jonathan (ed.), *Oxford Studies in the Philosophy of Religion*, Vol. 3 (New York: Oxford U.P., 2011).
Laplace, Pierre, *A Philosophical Essay on Probabilities* (Hong Kong: Forgotten Books, 2012 ed.).
Larson, Edward, *Evolution: The Remarkable History of a Scientific Theory* (New York: Modern Library, 2006).
Lennox, James, *Aristotle's Philosophy of Biology* (New York: Cambridge U.P., 2000).
Lenski, Richard and John Miller, "The Directed Mutation Controversy and Neo-Darwinism," *Science*, 259 (Jan. 1993), pp. 188–194.
Lewis, C.S., *The Problem of Pain* (New York: Macmillan, 1962).
Lewontin, Richard, *Biology as Ideology* (San Francisco, CA: HarperCollins, 1992).
Manson, Neil (ed.), *God and Design* (New York: Routledge, 2003).
Mayr, Ernst, *Toward a New Philosophy of Biology* (Cambridge, MA: Harvard, 1988).

Mayr, Ernst, *What Evolution Is* (New York: Basic, 2001).
McGrath, Alister, *Dawkins' God: Genes, Memes and the Meaning of Life* (Oxford: Blackwell, 2005).
McMullin, Ernan (ed.), *Evolution and Creation* (South Bend, IN: University of Notre Dame Press, 1985).
McMullin, Ernan, "Cosmic Purpose and the Contingency of Human Evolution," *Zygon*, 48.2 (June 2013), pp. 338–363.
McMullin, Ernan, "Evolution and Special Creation," *Zygon*, 28.3 (Sept. 1993), pp. 299–335.
McMullin, Ernan, "Plantinga's Defense of Special Creation," *Christian Scholar's Review*, XXI.1 (Sept. 1991), pp. 55–79.
Mellor, D.H., *Probability: A Philosophical Introduction* (London: Routledge, 2005).
Miller, Kenneth, *Finding Darwin's God* (San Francisco, CA: Harper, 2007).
Monod, Jacques, *Chance and Necessity* (New York: Knopf, 1971).
Moreland, J.P. and John Mark Reynolds (eds.), *Three Views on Creation and Evolution* (Grand Rapids, MI: Zondervan, 1999).
Morris, Simon Conway, *The Crucible of Creation* (New York: Oxford U.P., 2000).
Morris, Simon Conway, *Life's Solution* (New York: Cambridge, 2003).
Morris, Thomas (ed.), *Divine and Human Action* (Ithaca, NY: Cornell U.P., 1988).
Murphy, Nancey, *Beyond Liberalism and Fundamentalism* (Valley Forge, PA: Trinity Press, 1996).
Murphy, Nancey, *Bodies and Souls, or Spirited Bodies?* (New York: Cambridge U.P., 2006).
Murray, Michael, *Nature Red in Tooth and Claw* (New York: Oxford U.P., 2011).
Nagel, Thomas, *Mind and Cosmos* (New York: Oxford U.P., 2012).
Nasr, S.H., *The Need for a Sacred Science* (New York: SUNY Press, 1993).
Numbers, Ronald, *The Creationists* (Berkeley, CA: University of California Press, 1993).
Numbers, Ronald, *Darwinism Comes to America* (Cambridge, MA: Harvard U.P., 1998).
Okasha, Samir, *Philosophy of Science* (London: Oxford U.P., 2002).
Paley, William, *Natural Theology* (New York: Oxford U.P., 2010 ed.).
Palmer, Douglas, *Seven Million Years: The Story of Human Evolution* (London: Phoenix, 2006).
Peacocke, Arthur, *God and the New Biology* (San Francisco, CA: Harper, 1986).
Peacocke, Arthur, *Theology for a Scientific Age* (Minneapolis, MN: Fortress, 1993).
Pegis, Anton (ed.), *Introduction to St. Thomas Aquinas* (New York: Random House, 1945).
Pennock, Robert (ed.), *Intelligent Design Creationism and Its Critics* (New York: MIT Press, 2001).
Peterson, Michael, *God and Evil* (Boulder, CO: Westview, 1988).

Petrik, James, *Evil beyond Belief* (Armonk, NY: ME Sharpe, 2000).
Plantinga, Alvin, "When Faith and Reason Clash: Evolution and the Bible," *Christian Scholar's Review*, XXI.1 (Sept. 1991), pp. 8–32.
Plantinga, Alvin, *Where the Conflict Really Lies: Science, Religion and Naturalism* (New York: Oxford, 2011).
Polkinghorne, John, *Quantum Theory: A Very Short Introduction* (New York: Oxford, 2002).
Polkinghorne, John, *Science and Theology* (Minneapolis, MN: Fortress, 1998).
Popper, Karl, *Conjectures and Refutations* (London: Routledge, 2002).
Popper, Karl, *The Open Universe* (London: Routledge: 1982).
Richards, Janet Radcliffe, *Human Nature after Darwin* (London: Routledge, 2000).
Roberts, Jon H., *Darwinism and the Divine in America* (Madison, WI: University of Wisconsin Press, 1988).
Rolston, Holmes, *Science and Religion: A Critical Survey* (Philadelphia, PA: Templeton, 1987).
Rose, S., R.C. Lewontin, and L.J. Kamin, *Not in Our Genes: Biology, Ideology and Human Nature* (London: Penguin, 1990).
Rose, Steve (ed.), *The Richness of Life: The Essential Stephen J. Gould* (New York: Norton, 2007).
Rowe, William, *Philosophy of Religion* (Belmont, CA: Wadsworth, 2007, 4th ed.).
Ruelle, David, *Chance and Chaos* (Princeton, NJ: Princeton U.P., 1991).
Ruse, Michael, *Darwin and Design* (Cambridge, MA: Harvard U.P., 2004).
Ruse, Michael, *Darwinism and Its Discontents* (New York: Cambridge U.P., 2006).
Ruse, Michael, *Monad to Man* (Cambridge, MA: Harvard U.P., 1997).
Ruse, Michael, *Taking Darwin Seriously* (Amherst, NY: Prometheus, 1998).
Russell, Robert, *et al.* (eds.), *Chaos and Complexity* (Vatican City: Vatican Observatory, 2000).
Russell, Robert, *et al.* (eds.), *Evolutionary and Molecular Biology* (Vatican City: Vatican Observatory, 1998).
Russell, Robert, *et al.* (eds.) *John Paul II on Science and Religion* (Vatican City: Vatican Observatory, 1990).
Sagan, Carl, *Billions and Billions* (New York: Ballantine, 1997).
Sagan, Carl, *Cosmos* (New York: Random House, 2002).
Sagan, Carl, *Dragons of Eden* (New York: Ballantine, 1986).
Saunders, Nicholas, *Divine Action in the World* (Cambridge: Cambridge U.P., 2002).
Schőnborn, Christoph, *Chance or Purpose*? (San Francisco, CA: Ignatius, 2007).
Searle, John, *Freedom and Neurobiology* (New York: Columbia U.P., 2006).
Shanahan, Timothy, *The Evolution of Darwinism* (New York: Cambridge, 2004).
Shanahan, Timothy, "The Evolutionary Indeterminism Thesis," *BioScience*, 53.2 (Feb. 2003), pp. 163–169.
Simpson, George, *The Meaning of Evolution* (New Haven, CT: Yale U.P., 1949).

Simpson, George, *This View of Life* (Fort Worth, TX: Harcourt, 1966).
Sober, Elliott, *Evidence and Evolution* (Cambridge: Cambridge U.P., 2008).
Sweetman, Brendan, *Religion and Science: An Introduction* (New York: Continuum, 2010).
Sweetman, Brendan, *Religion: Key Concepts in Philosophy* (New York: Continuum, 2007).
Sweetman, Brendan (ed.), *Philosophical Thinking and the Religious Context* (New York: Bloomsbury, 2013).
Sweetman, Brendan, "The Dispute between Plantinga and McMullin over Evolution," *American Catholic Philosophical Quarterly*, 86.2 (Spring 2012), pp. 343–354.
Sweetman, Brendan, "Evolution, Chance and Necessity in the Universe," *Portuguese Journal of Philosophy*, 66.4 (2010), pp. 897–910.
Swinburne, Richard, *Is There a God*? (New York: Oxford U.P., 1996).
Swinburne, Richard, *The Existence of God* (New York: Oxford U.P., 2004).
Taylor, Richard, *Metaphysics* (Englewood Cliffs, NJ: Prentice Hall, 1963).
Teilhard de Chardin, Pierre, *The Phenomenon of Man* (San Francisco, CA: Harper, 2008).
Van Inwagen, Peter, *God, Knowledge and Mystery* (Ithaca, NY: Cornell, 1995).
Van Till, Howard, *The Fourth Day* (Grand Rapids, MI: Eerdmans, 1986).
Wallace, Alfred, R., *Contributions to the Theory of Natural Selection* (London: Macmillan, 1870).
Ward, Keith, *God, Chance and Necessity* (Oxford: Oneworld, 1996).
Weinberg, Steven, *The First Three Minutes* (New York: Basic Books, 1993).
Wilson, Edward O., *Sociobiology: The New Synthesis* (Cambridge, MA: Harvard U.P., 1975).

Index